Lecture Notes in Computer Science   14970

Founding Editors

Gerhard Goos
Juris Hartmanis

Editorial Board Members

Elisa Bertino, *Purdue University, West Lafayette, USA*
Wen Gao, *Peking University, Beijing, China*
Bernhard Steffen, *TU Dortmund University, Dortmund, Germany*
Moti Yung, *Columbia University, New York, USA*

The series Lecture Notes in Computer Science (LNCS), including its subseries Lecture Notes in Artificial Intelligence (LNAI) and Lecture Notes in Bioinformatics (LNBI), has established itself as a medium for the publication of new developments in computer science and information technology research, teaching, and education.

LNCS enjoys close cooperation with the computer science R & D community, the series counts many renowned academics among its volume editors and paper authors, and collaborates with prestigious societies. Its mission is to serve this international community by providing an invaluable service, mainly focused on the publication of conference and workshop proceedings and postproceedings. LNCS commenced publication in 1973.

Abdelkader Hameurlain · A Min Tjoa
Editors

Transactions on Large-Scale Data- and Knowledge-Centered Systems LVII

 Springer

Editors-in-Chief
Abdelkader Hameurlain
IRIT, Paul Sabatier University
Toulouse, France

A Min Tjoa
IFS, Vienna University of Technology
Vienna, Austria

ISSN 0302-9743 ISSN 1611-3349 (electronic)
Lecture Notes in Computer Science
ISSN 1869-1994 ISSN 2510-4942 (electronic)
Transactions on Large-Scale Data- and Knowledge-Centered Systems
ISBN 978-3-662-70142-3 ISBN 978-3-662-70140-9 (eBook)
https://doi.org/10.1007/978-3-662-70140-9

© The Editor(s) (if applicable) and The Author(s), under exclusive license
to Springer-Verlag GmbH, DE, part of Springer Nature 2025

This work is subject to copyright. All rights are solely and exclusively licensed by the Publisher, whether the whole or part of the material is concerned, specifically the rights of translation, reprinting, reuse of illustrations, recitation, broadcasting, reproduction on microfilms or in any other physical way, and transmission or information storage and retrieval, electronic adaptation, computer software, or by similar or dissimilar methodology now known or hereafter developed.
The use of general descriptive names, registered names, trademarks, service marks, etc. in this publication does not imply, even in the absence of a specific statement, that such names are exempt from the relevant protective laws and regulations and therefore free for general use.
The publisher, the authors and the editors are safe to assume that the advice and information in this book are believed to be true and accurate at the date of publication. Neither the publisher nor the authors or the editors give a warranty, expressed or implied, with respect to the material contained herein or for any errors or omissions that may have been made. The publisher remains neutral with regard to jurisdictional claims in published maps and institutional affiliations.

This Springer imprint is published by the registered company Springer-Verlag GmbH, DE,
part of Springer Nature
The registered company address is: Heidelberger Platz 3, 14197 Berlin, Germany

If disposing of this product, please recycle the paper.

Preface

This volume contains five fully revised regular papers, covering a wide range of very hot topics focused on leveraging machine learning for effective data management, access control models, reciprocal authorizations, Internet of Things, digital forensics, code similarity search, volunteered geographic information, and spatial data quality.

We would like to sincerely thank the editorial board for thoroughly refereeing the submitted papers and ensuring the high quality of this volume. In addition, we would like to express our wholehearted thanks to the team at Springer for their ready availability, the efficiency of their management and the very pleasant cooperation in the realization of the TLDKS journal volumes.

June 2024

Abdelkader Hameurlain
A Min Tjoa

Organization

Editors-in-Chief

Abdelkader Hameurlain   Paul Sabatier University, IRIT, France
A Min Tjoa   Technical University of Vienna, IFS, Austria

Editorial Board

Reza Akbarinia	Inria, France
Dagmar Auer	Johannes Kepler University Linz, Austria
Djamal Benslimane	University Lyon 1, France
Stéphane Bressan	National University of Singapore, Singapore
Mirel Cosulschi	University of Craiova, Romania
Johann Eder	Alpen Adria University of Klagenfurt, Austria
Anna Formica	National Research Council in Rome, Italy
Shahram Ghandeharizadeh	University of Southern California, USA
Anastasios Gounaris	Aristotle University of Thessaloniki, Greece
Sergio Ilarri	University of Zaragoza, Spain
Petar Jovanovic	Universitat Politècnica de Catalunya and BarcelonaTech, Spain
Aida Kamišalić Latifić	University of Maribor, Slovenia
Dieter Kranzlmüller	Ludwig-Maximilians-Universität München, Germany
Philippe Lamarre	INSA Lyon, France
Lenka Lhotská	Technical University of Prague, Czech Republic
Vladimir Marik	Technical University of Prague, Czech Republic
Jorge Martinez Gil	Software Competence Center Hagenberg, Austria
Riad Mokadem	Paul Sabatier University, IRIT, France
Franck Morvan	Paul Sabatier University, IRIT, France
Torben Bach Pedersen	Aalborg University, Denmark
Günther Pernul	University of Regensburg, Germany
Viera Rozinajova	Kempelen Institute of Intelligent Technologies, Slovakia
Soror Sahri	LIPADE, Université Paris Cité, France
Joseph Vella	University of Malta, Malta
Shaoyi Yin	Paul Sabatier University, IRIT, France
Feng "George" Yu	Youngstown State University, USA

Contents

Leveraging Machine Learning for Effective Data Management 1
   Sana Sellami

Exploring Reciprocal Exchanges and Trust-Based Authorizations:
A Feasibility Demonstration with Location-Based Services 27
   Gabriela Suntaxi, Aboubakr Achraf El Ghazi, and Klemens Böhm

Device Forensics in Smart Homes: Insights on Advances, Challenges
and Future Directions 68
   Sabrina Friedl and Günther Pernul

Evaluation of Code Similarity Search Strategies in Large-Scale Codebases 99
   Jorge Martinez-Gil and Shaoyi Yin

Quality Assessment of Volunteered Geographic Information: A Survey 114
   Donia Nciri, Salma Sassi, Richard Chbeir, and Sami Faiz

Author Index 151

Leveraging Machine Learning for Effective Data Management

Sana Sellami$^{(\boxtimes)}$ [iD]

Aix Marseille Univ, CNRS, LIS, Marseille, France
sana.sellami@univ-amu.fr

Abstract. The exponential growth of heterogeneous data from diverse sources, such as social media, IoT sensors, and transactional databases, poses significant challenges for effective processing and analysis. This data, often characterized by poor quality, diverse formats, and complex structures, prevents its utilization for extracting valuable insights and supporting informed decision-making. Machine learning (ML) emerges as a powerful tool to address these challenges by automating heterogeneous data processing tasks and enhancing data quality, integration, and analysis.

In this context, this paper explores the contribution of machine learning methods to the different stages of the data management process: preparation, integration, and analytics. We aim to provide a comprehensive study of the role these methods play throughout the entire pipeline, as well as highlighting a set of challenges in this field.

Keywords: Data Management · Machine Learning · Data Quality · Data Preparation · Data Integration · Data Analytics · Big Data

1 Introduction

Data plays a crucial role in decision-making processes. The increasing proliferation of data, known as Big Data [64], presents several challenges related to its processing, including volume, variety, velocity, quality, and integration. Originating from diverse sources, such as the Web and sensors in connected environments, this data is heterogeneous, as it can be structured, semi-structured (e.g., JSON, XML), or unstructured (e.g., text). Additionally, it varies significantly in terms of syntax, structure and semantics.

A major problem with this heterogeneous data is its lack of quality, which manifests in issues such as missing, redundant, or outlier data, making it complex to exploit. Therefore, it is essential to preprocess the data to enhance its quality, integrate it, and make it potentially exploitable.

In this context, Big Data Management (BDM) [91] is a discipline that encompasses all the methods and tools for collecting, modeling, integrating, analyzing, and using data throughout all stages of their lifecycle, from their creation to

---

© The Author(s), under exclusive license to Springer-Verlag GmbH, DE, part of Springer Nature 2025
A. Hameurlain and A. M. Tjoa (Eds.): *Transactions on Large-Scale Data- and Knowledge-Centered Systems LVII*, LNCS 14970, pp. 1–26, 2025.
https://doi.org/10.1007/978-3-662-70140-9_1

their deletion. It has become essential for several businesses, industries, and in various application domains.

Figure 1 illustrates the generic process of BDM, which includes:

Fig. 1. Big Data Management Process

- **Data Preparation:** This is the process of transforming raw data into structured and cleaned data to make it ready for analysis [60]. In data warehouses, data preparation is an ETL (Extract-Transform-Load) process [100] applied to the data. For ad hoc analyses conducted by data scientists, data preparation involves a series of tasks collectively known as data wrangling [82]. These tasks include data profiling, which aims to understand the structure and content of the data, data validation, data cleaning to detect and correct errors and data transformation.
- **Data Integration:** This relies on a key step called Entity Resolution (ER) [77], to identify different descriptions that refer to the same real-world entities. Entity resolution is based on Matching, which is the process of determining correspondences between structured (e.g., relational data) or semi-structured (e.g., XML, RDF) data. Traditionally, matching has been used to determine correspondences between pairs of attributes from a source schema and a target schema [14]. Today, it represents a fundamental component for determining semantically similar attributes [12,80] and classifying relationships between datasets [61]. It is used in various applications such as named entity recognition, the Web, recommendation systems, etc.
- **Data Analytics** [98]: It represents a set of processes for cleaning, transforming, and modeling data to discover and produce new information that could improve decision-making.

To meet these needs, particularly those related to data preparation, integration, and analytics, the literature offers a wide range of methods, from rule-based methods to machine learning (ML) models. Indeed, for more than a decade, we have witnessed the emergence of ML techniques, which have become indispensable tools for automating these processes [16,63,72,83,102].

In this paper, we illustrate the contribution of machine learning methods to various stages of the data management process. Specifically, we address how these solutions tackle issues related to data quality, heterogeneity, non-stationarity, and imbalanced data. We then present a comprehensive study of the role these methods play throughout the entire pipeline, highlighting key research directions in this field.

The rest of this paper is structured as follows. Section 2 provides a description of the research challenges. Section 3 reviews works on data preparation, particularly data imputation and anomaly detection. In Sect. 4, we review data integration tasks and the important role that machine learning (ML) plays in these tasks. Section 5 addresses data analytics and the challenges related to imbalanced data and trustworthiness. Finally, Sect. 6 concludes this paper and highlights future directions.

2 Challenges

The data management process lies at the heart of effectively managing information from various sources. However, this process often faces significant challenges.

Firstly, data quality is an essential aspect to take into account. Indeed, poor data quality can lead to difficulties in their processing and analysis [10]. Next, given the diversity of data sources, data integration is of central importance to reconcile this varied information, make it compatible, and merge it coherently. Finally, data imbalance can impact the data analytics process. Some categories of data may be overrepresented, while others are underrepresented, which can bias the analysis results.

Quality: Data quality can be assessed across different dimensions [92], such as completeness to identify missing values, reliability to ensure accuracy, error-free, and consistency, and uniqueness to detect duplicates in the dataset. *Rahm and Do* [81] proposed a classification of different types of errors that can occur in an ETL (Extract-Transform-Load) process, which can be: single-source or multi-source, schema-level or instance-level. Schema-level issues are related to the structure of the data, the schema that defines how the data is organized. They can be resolved through schema transformation, evolution, and integration. In contrast, instance-level issues refer to errors and inconsistencies within the data content, emphasizing the importance of data cleaning.

It is important to note that these data quality issues closely depend on the type of data and the context in which they are used. Problems related to a single source also occur in the case of multiple sources. At the schema level, these issues are related to data heterogeneity (e.g., data models), the presence of naming conflicts (e.g., homonyms, synonyms), or structural conflicts (e.g., different data types, different integrity constraints). At the instance level, one can mention issues of data inconsistency, contradiction, etc.

Heterogeneity: Several data management applications rely on connections (links) between data to better discover them. These applications require the integration of different data referring to the same real-world object. Data, even within the same domain, may exhibit a diversity of schemas and formats, creating the necessity to align these descriptions to facilitate their discovery and analysis. Indeed, they are often represented in different structures (structured, semi-structured, or unstructured) and extracted from different sources (e.g., Web, databases) that can be stored locally or distributed. Moreover, data from a single source may have different schemas. Resolving these data conflicts improves the performance of the integration process.

Stationarity and Imbalance: Non-stationary data refers to data where statistical properties change over time. This can manifest as shifts in the underlying distribution or changes in the concept that the data represents, known as conceptual drift. Imbalance, antoher challenge, refers to a difference between classes in classification tasks or the number of instances in regression tasks. Both challenges are particularly relevant for streaming data collected in real-time, as its dynamic nature can lead to distribution changes.

Furthermore, if the imbalance leads to conceptual drift, it can result in loss of accuracy and model performance, thus impacting data analysis. Traditional algorithms, designed to discover knowledge from data, assuming that data properties are fixed, will be limited in terms of performance, adaptability, and robustness to address them [3]. Therefore, we need models/algorithms capable of adapting to conceptual drift, such as those employing explicit detectors or implicit adaptation mechanisms.

3 Data Preparation

Data preparation is a set of operations aimed at detecting errors and improving data quality. It represents a key step as the reliability of the analysis largely depends on data quality. Data preparation involves several operations [46], among which:

- **Data discovery** is the process of analyzing and collecting data from different sources to identify missing data or detect outliers.
- **Data validation** involves applying rules and constraints to analyze the data and determine, for instance, if it is correct or complete.
- **Data structuring** includes tasks for creating, representing, and organizing information and transforming data.
- **Data enrichment** enhances data with new values or derived values, such as inserting metadata or generating primary keys.
- **Data cleaning** allows for the removal, addition, or replacement of imprecise values with more accurate and representative values.

More specifically, we will focus in this paper on data cleaning [81] which is an essential step when integrating heterogeneous data sources [32]. It involves two phases [32]: 1) *error detection* such as duplicates, missing values, constraint violations and incorrect values, and 2) *error repair*, which includes updating available data to correct or remove identified errors.

Two kind of approaches for error detection have been proposed in the literature: 1) qualitative approaches [31] which rely on descriptive methods to specify patterns or constraints of a data instance and thus identify data that violate these patterns or constraints, and 2) quantitative approaches [17] that use statistics and other analytical techniques to detect, quantify and correct quality problems. On the other hand, error repair is performed by applying either data transformation scripts, which are often generated according to the process used for error detection, or involving human expertise, or by combining these two strategies.

Traditionally, cleaning is performed on structured data using rule-based approaches where the goal is to check if integrity, denial, and functional constraints are satisfied [32]. However, with the proliferation of digitization in various fields of application (such as health, transportation, or smart cities), we face new challenges as the collected data are unstructured, distributed, produced in real-time, and highly heterogeneous. In this context, rule-based approaches become almost obsolete given the explosion of the number of rules in data profiling and the complexity of managing the consistency of these rules [51].

ML for Data Cleaning: The emergence of new data cleaning methods based on machine learning (ML) has addressed the limitations of traditional approaches [51]. Indeed, the use of ML has automated some tasks such as deduplication, anomaly or outlier detection. Several recent works have proposed systems and approaches for data detection and repair using mainly machine learning techniques. Among them is HoloClean [85], which is a data cleaning system that allows the repair of structured data by relying on the use of integrity constraints, external knowledge, and quantitative statistics. It exploits machine learning to validate corrections using a probabilistic graphical model. AlphaClean [62] is a framework for data cleaning that uses machine learning to automatically generate and adjust data cleaning pipelines based on quality measures specified by the user. AlphaClean offers a common intermediate representation for data repair to optimize the data cleaning process. The Learn2Clean method [15] is based on Q-Learning, a reinforcement learning technique that selects, for a specific dataset, an ML model and a quality metric, an optimal sequence of data preprocessing tasks such that the quality, in terms of performance, of the resulting ML model is maximized.

The following sections will explore the key challenges in data cleaning, specifically addressing issues like missing values and anomalies.

3.1 Data Imputation

According to [93], there are several methods for handling missing values that operate based on one of the following actions:

1. Deleting incomplete observations;
2. Manually repairing (i.e., human experts use domain expertise to fill in the missing values);
3. Replacing with a constant / the last observation / the mean;
4. Estimating the most probable value, a method known as *Imputation*.

Data Imputation uses the data from collected observations as much as possible to estimate the missing values [93], preserves data that are only partially missing (e.g., a vector containing one or more missing elements), and does not require human intervention.

Several methods have been proposed in the literature for data imputation, such as constraint-based methods using, for example, Petri nets, statistical models, clustering (e.g., K-means [39]), regression and autoregressive models (e.g., ARMA (AutoRegressive Moving Average), ARIMA (AutoRegressive Integrated Moving Average)). Recently, deep learning-based methods have been proposed [94], such as recurrent neural networks (RNN) [99] and their variants like LSTM (Long Short Term Memory) and GRU (Gated Recurrent Unit), as well as non-autoregressive deep generative models to predict missing values.

For streaming time series data, various approaches have been proposed for online imputation [57,79,103] that can be grouped into two categories: matrix completion techniques and pattern matching techniques.

Matrix completion techniques assume temporal continuity in the data streams and use matrix decomposition or factorization methods, such as Principal Component Analysis (PCA) [8]. However, these methods rely on the assumption that the data is static and do not exploit the temporal dimension of the data, i.e., the evolution of the data over time [105].

Pattern matching techniques aim to identify recurrent patterns within the data streams and then exploit them to estimate missing data [6]. These methods operate in real-time, analyzing dependencies between different data streams. However, finding the complete set of exact dependencies between the data may require significant resources, potentially hindering the efficiency of the imputation process. Pattern matching techniques distinguish themselves from matrix completion methods by their ability to accurately capture temporal variations. Among these techniques, some integrate deep learning methods due to their high performance [105]. However, they face a major challenge: concept drift. Indeed, these methods require large volumes of representative data and significant computational resources to adapt to new concepts, limiting their relevance in environments prone to frequent changes [18,109].

3.2 Anomaly Detection

Anomaly detection is the problem of finding unusual observations and patterns in data. These deviations in data is often referred to the terms of anomalies or outliers, which can be used interchangeably.

Choosing the right technique to find anomalies depends on the type of anomaly that should be detected. There are three different categories in which anomalies can be classified [27]:

– **Point Anomaly:** A single instance standing out with respect to other data can be defined as point anomaly.
– **Contextual Anomaly:** A single instance deviating from an expected pattern, i.e. a specific context, can be defined as contextual anomaly.
– **Collective Anomaly:** A collection of observations deviating from common patterns is defined as a collective anomaly.

The idea of detecting abnormal observations has attracted many researchers [27] and companies to develop algorithms for various use cases such as intrusion detection, fraud detection, Medical Health Anomaly Detection, sensor networks, etc. In fact, it has the capabilities of providing more autonomy, reliability or security to applications.

An array of methodologies has been employed for anomaly detection, ranging from traditional statistical techniques to state-of-the-art deep learning approaches. Statistical methods are among the computationally most efficient while delivering good results [22]. They work by capturing statistical properties of the data and detect anomaly based on historical data. Several algorithms have caught attention [27] as Autoregressive Model, Moving Average Model. Autoregressive Moving Average combines these two ideas and Autoregressive Moving Integrated Average generalizes the idea by relaxing the data stationarity assumption. Many machine learning techniques have been used to find outliers in data. While some of them were developed for different tasks, they have also proven to be useful for time series data. Among most popular are clustering-based algorithms which aim to group similar instances into clusters, e.g. DBSCAN [38].

Unlike traditional machine learning approaches, deep neural networks excel in handling complexity and flexibility enabling them to scale to high dimensional data. Indeed, with the rise of deep neural networks various new methods have been found to approach anomaly detection. They have shown advancements in lots of domains, among them working with time series data [21]. Their popularity is based on empirical results being able to regularly outperform previous models. In contrast to traditional machine learning approaches, deep neural networks overcome their limitation of complexity and flexibility. This restricts their ability to scale to high dimensional data which is present frequently nowadays [97].

4 Data Integration

Data integration is the process of bringing together data from various sources (e.g., databases, applications) to create a unified and comprehensive view of the data. It comprises different tasks [35]: data extraction, schema matching, entity resolution/matching and data fusion.

In the literature, we distinguish between works focusing on schema matching and ontology matching [34,90]. Ontology matching aims to determine semantic correspondences between entities belonging to different ontologies. Although there is a difference between schema matching and ontology matching problems [89], techniques developed for each of these domains can mutually benefit from one another.

Entity matching is a critical step in data integration because it ensures that entity descriptions refer to the same real-world entity. Traditionally, entity matching relies on the use of similarity measures to identify correspondences between attributes of different schemas. However, these approaches, although interpretable, often involve costly human intervention to adjust parameters and reduce the risk of false positives [88]. Moreover, they do not fully capture the semantics of the data.

In complex data integration pipelines, matching may go beyond simple similarity measurement between two data sources. It may also encompass noisy data and variations, such as different names or specialization and generalization relationships. To address these challenges, it is essential to employ more advanced methods. In this regard, machine learning and deep learning models emerge as promising solutions for resolving data matching and qualification problems [9].

ML for Data Integration: Recently, ML methods have been applied to better capture semantic relationships and determine alignment between entities [35]. The recent advancements in machine learning have brought significant progress in the field. The study proposed in [35] describes the synergy between data integration and ML methods in various integration tasks. For entity matching, early works utilized probabilistic models that determine attribute similarities as dimensions of comparison vectors, each representing the probability that each pair of descriptions is similar [29]. Subsequently, several machine learning models have been employed [9]: supervised, unsupervised, and more recently, deep learning.

To better understand the roles these methods play in data integration, we will explore three tasks: schema matching, entity matching, and data fusion, which are crucial aspects of the integration process.

4.1 Schema Matching

Schema matching approaches are generally divided into two categories [12, 14, 80]: schema-based and instance-based. Hybrid/composite approaches combine these two categories to improve accuracy. Several techniques have been used in this context, including [14]: 1) linguistic-based matching relies on the name or description of an element, utilizing stemming, tokenization, string matching, and information extraction techniques, 2) the use of auxiliary information such as thesauri or dictionaries, 3) instance-based matching means schema elements are considered similar if their instances are similar, relying on statistics, metadata, or trained classifiers, and 4) structural matching considers schema elements as similar if they appear in similar structural groups, have similar relationships, or have connections (paths) to other similar elements. Finally, some schema matching approaches integrate user feedback, which includes tuning parameters, answering questions, and reusing previous resources.

It's common to consider schema matching as an essential preliminary step before performing entity matching [9]. However, schema matching can also be an integral part of the entity matching process. Indeed, some methods, especially those based on deep learning, are capable of handling schema matching jointly with other steps of the process. Moreover, in many real-world situations, schema matching is essential and cannot be overlooked. Therefore, in practice, this step is often integrated into data preprocessing to harmonize schemas and facilitate comparison between data sources [9].

4.2 Entity Matching

Entity Matching aims to identify different descriptions that refer to the same entities in the real world and is fundamental for data integration and cleaning problems. In the literature, entity matching, entity resolution, and data linkage are frequently used to address the same problem. According to Papadakis et al. [76], entity matching methods have evolved over the years into four generations, each addressing challenges posed by the four Vs: Veracity (1st generation), Volume (2nd generation), Variety (3rd generation), and Velocity (4th generation).

These generations continue to play a crucial role in current research [76], with interest in using deep learning and crowdsourcing to improve performance across all stages of the entity matching process.

Indeed, the methods for entity matching have seen significant advancement in recent years, as illustrated in Fig. 2 inspired by [76]. Early entity matching works relied on rule-based methods using similarity measures between strings. With the rise of machine learning, several approaches based on probabilistic methods, supervised, unsupervised and deep learning have been proposed in the literature [9, 30].

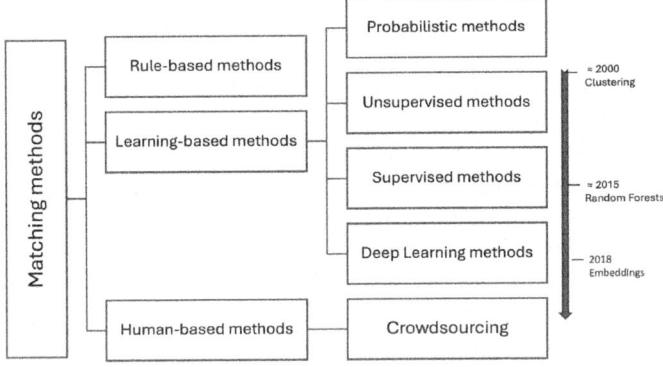

Fig. 2. Evolution of matching methods

Rule-Based Methods: Rule-based matching involves determining correspondences between two entities using similarity measures. If the similarity exceeds a certain threshold $\in [0, 1]$, then these two entities are considered similar. Several studies have explored these measures[1] [33], which can be either: Measures based on character sequences, Token-based measures, Hybrid approaches combining the characteristics of the two previous methods, Phonetic measures, and Domain-dependent measures.

---

[1] https://hpi.de/naumann/teaching/teaching/ss13/data-profiling-and-data-cleansing.html.

Machine Learning Based Methods: These methods compute similarities between attributes and use them as features. Recent works [24, 29] have demonstrated that ML-based techniques achieve much higher accuracies than rule-based approaches.

In fact, ML-based methods consider matching as a classification problem. They are attribute-based because they mainly attempt to learn the probability that two descriptions match based on previous examples of similar pairs, as their models are trained on sets of pairs or vector representations of entity descriptions or words used in the values of these descriptions.

Supervised Learning: Supervised learning approaches for entity matching emerged more than two decades ago and utilized decision trees, logistic regression, and SVM [35]. Recent ML models such as random forests have significantly improved matching. One of the works that made a significant breakthrough in supervised learning is the *Magellan* project [59], which covers the entire processing pipeline including blocking and matching of structured data, as well as sampling and debugging.

Unsupervised Learning: Unsupervised learning does not require labeled data. We cite the work of Papadaki et al. [78], who proposed the use of unsupervised learning on structured and semi-structured data and developed an entity matching system named *Jedai*.

Magellan and Jedai are two examples of entity matching systems that integrate both string similarity measurement techniques and learning algorithms, whether supervised or unsupervised, within their entity matching process.

These systems are particularly effective when applied to structured entities with short attributes. However, they face two challenges: (1) the inability to consider semantic similarity between entities, and (2) their sensitivity to noisy and textual data. This is partly due to the fact that they have few significant features to exploit using string similarity measures.

Moreover, one of the limitations of supervised approaches is the need for a sufficient dataset for model training, as an uninformative or unrepresentative dataset can lead to overfitting or biased and inaccurate classifiers. The problem lies in the cost of generating a labeled training set. This involves manually labeling a quantity of examples because no matches are much more numerous than matches, requiring considerable human effort. The emergence of weak supervision methods, automatic self-learning models [42], active learning [23, 28, 52], transfer learning [96], or self-supervised learning [42] has partially addressed the problem of generating a labeled dataset. Indeed, these approaches are promising when labeled examples are available or easy to generate using existing tools and when test data does not deviate significantly from training data.

Active Learning: The objective of active learning is to reduce the number of pairs that need to be labeled by selecting the most representative examples. It can be considered as a labeling protocol acting as semi-supervised learning by choosing which data to label. The choice of instances to label for learning can significantly influence the quality of the learned classifier. The use of active learning for entity matching is not new [13], and most work has focused on selecting pairs based on a classification model. Active learning is an iterative process where, in each iteration, a number of representative and unlabeled instances are selected and manually labeled by a human oracle.

Deep Learning: Recent advancements in deep learning have greatly influenced research in entity resolution and matching [9,30]. Unlike entity matching solutions based on learning methods such as random forests or SVM, deep learning methods are able of learning from raw text, as demonstrated in tasks related to natural language processing (NLP). Deep learning enables the comparison of values from long texts using their vector representations (embedding) and represents a promising approach for matching unstructured (e.g., text) and dirty data. Most deep learning methods used in the matching process integrate feature extraction, record pair comparison, and classification into a single step like a neural network [9]. The most commonly used methods include recurrent neural networks (RNNs), word embeddings, and recently, transformer-based pretrained language models [24,108].

Crowdsourcing: It is a type of outsourcing that leverages the collective knowledge or ideas of individuals. The assumption is that humans can improve matching accuracy by leveraging contextual information and common sense [29]. This technique has been applied and exploited in several research domains (e.g., pattern recognition, ontology alignment, semantic annotation) and for data management [67]. The crowd is solicited to perform online tasks called Human Intelligence Tasks (HITs). Users are often better than algorithms at detecting different terms that refer to the same entity. However, compared to algorithmic techniques, users are much slower and more expensive. Some works have proposed a hybrid user-machine approach for generating HITs, combining the efficiency of machine-oriented approaches with the quality of responses obtained from the crowd [29,101]. Recently, new approaches for reducing uncertainty in matching using Crowdsourcing have been proposed. We mention the work of [107], which proposed two new approaches to select and publish an optimal set of questions based on new responses received.

However, although Crowdsourcing has improved matching for verification, validation, and uncertainty reduction, it remains a financially costly process and can only be used by domain experts. Other user-oriented approaches (Human Schema Matching) [88] have proposed integrating humans into the entire matching process. However, a recent study [2] demonstrated that humans have cognitive biases that reduce their ability to effectively perform tasks.

Synthesis: Recent approaches to entity matching primarily rely on two categories of models: (1) traditional machine learning (ML) models and (2) deep learning (DL) models.

Approaches in the first category are effective mainly for syntactic matching tasks involving structured or semi-structured data with little noise. However, they do not consider semantic similarity between entities. This limitation arises from the nature of the features used, which rely on similarity measures between strings. Yet, considering semantic similarity is crucial for identifying entity pairs with various variations, such as different names, synonyms, or polysemies. DL approaches have addressed this limitation by calculating similarity between vector representations of words (static or contextual). Moreover, these approaches perform well even in the presence of heavily noisy and unstructured data.

Finally, it's worth noting that one of the limitations of machine learning-based methods is that matches can sometimes be invalid. Consequently, the matching process remains complex, unable to guarantee 100% accuracy. Hence, the proposal to integrate a step of human verification and validation into this automated process, using crowdsourcing methods, as described in the work of Wang et al. [101].

4.3 Data Fusion

Data fusion [26] is the process of combining data from diverse sources to create information of higher quality and relevance than the raw data, which may be uncertain, imprecise, inconsistent, and contradictory. Data fusion plays an important role in data integration systems because it can detect and remove noisy data and increase the accuracy of the integrated data [36].

There are three levels of data fusion [95]: early fusion, intermediate fusion, and late fusion. According to Stahlschmidt et al. [95], the performance of these different strategies closely depends on the problem and data characteristics.

In early fusion, input data is concatenated, and the resulting vector is treated as a unimodal input. Thus, the model does not differentiate the origins of the modality features. This approach is typically performed on homogeneous modalities, i.e., when there is only one type of modality. It is advantageous for its simplicity but does not allow for identifying relationships between modalities.

In intermediate fusion, feature representations in the form of feature vectors are learned by models and then fused. This type of fusion is capable of learning complex correlations between features from different modalities.

In late fusion, instead of combining data or features, the decisions of models for each modality are combined to produce a final decision. This allows for better learning as each model can be tailored to a specific modality. However, this approach does not consider correlations at the feature level.

5 Data Analytics

Data analytics aims to extract information to assist in making predictions, identifying recent trends, and improving decision-making.

With the advent of Big Data, we have been facing massive and heterogeneous data, posing new challenges in their processing and analysis [71]. Techniques for analyzing this massive data are known as Big Data Analytics (BDA) [98]. This refers to the process of collecting, organizing, and analyzing data [4] (Fig. 3) with the four aspects (the 4 Vs) of big data: Volume, Velocity, Variety, and Veracity.

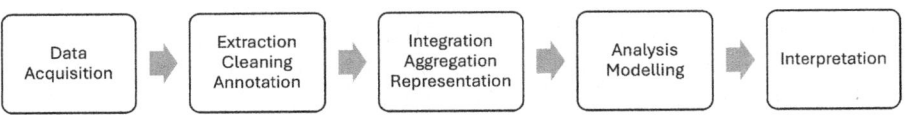

Fig. 3. Big data analytics pipeline

The analytical process must therefore take into account these different specificities [65,98]:

- From a volume perspective, the deluge of input data is the initial aspect to consider, as it can impact data analysis, primarily due to challenges associated with data imbalance, non-linearity, or the curse of dimensionality.
- From a velocity perspective, real-time or streaming data raises the challenge of processing a large amount of data arriving within a limited time. For instance, data from connected objects not only require real-time processing but also consideration of these objects' processing capacity. Algorithms must be developed to account for conceptual drift, be deployable, and have optimized computational capabilities.
- From a variety perspective, data can be of different types or incomplete, necessitating the development of analysis models capable of handling heterogeneity and issues related to data quality such as incompleteness and noisy and erroneous data.
- From a veracity perspective, it is necessary to consider data provenance and uncertainty. Veracity refers not only to data reliability but also to the lack of reliability in data sources. Data provenance is the process that allows traceability to identify error sources. Evaluating data uncertainty is a crucial step in the BDA process [47]. Although there are various techniques for analyzing massive data, analysis accuracy can be negatively affected by data uncertainty.

Data analytics requires tools and methods such as data mining, which plays a significant role in massive data analysis in terms of computational cost, memory, and result accuracy. Several methods can be used [98], such as classification algorithms, clustering, or frequent pattern extraction. Data mining algorithms are designed for specific problems. Unlike these algorithms, machine learning models can be used for various data mining and analysis problems as they are used as search algorithms for a certain solution.

Machine Learning for Data Analytics: Machine learning models have become a principal component in massive data analytics [65]. This is due to their ability to learn from data, extract knowledge, and provide predictions for better decision-making. Depending on the nature of the data, there are two categories of learning tasks: supervised when input data and outputs (labels) are known, and unsupervised when the model must discover the structure of the data itself. Classification models and regression models are examples of supervised learning. In classification, data are discrete values, while regression predicts a number based on historical observations and probabilities.

The study proposed in [75], discusses the ML techniques used for big data analytics. This study reveals that the most commonly used models are SVM models, decision trees due to their simplicity, and deep learning models [45] such as LSTM (Long Short-Term Memory) and convolutional neural networks. Other algorithms are also widely used, such as Bayesian inference, gradient boosting, and ensemble analysis models. These methods rely on data for decision-making.

Furthermore, the literature offers a rich array of works describing the analytics process from both a general perspective [4,54] and domain-specific applications, such as healthcare [11] and the Internet of Things [72].

Despite these advancements, the field faces ongoing challenges including the real-time processing of streaming data, managing imbalanced datasets, and ensuring the reliability of predictions. These challenges underscore the critical need for further advancements and solutions in ML techniques designed for handling them.

5.1 Learning from Stream Data

Data streams are defined [7] as "unlimited sequences of multi-dimensional, irregular, and temporary observations available over time" and have the following characteristics [3]: being voluminous, infinite, non-stationary, uncertain, or having low quality due, for example, to the presence of missing values.

Managing data streams represents a significant challenge due to the massive volume and the speed at which information is generated, as well as the conceptual drift caused by the non-stationarity of data [7]. To address these issues, algorithms must be capable of incrementally processing incoming information [7]. Incremental learning corresponds to a system able of receiving and integrating new examples without needing to perform complete retraining.

Data streams come in two main forms [3]: online, where instances are provided sequentially, and as chunks or blocks of data. Two essential approaches emerge to learn form these data: online learning and chunk-based learning.

Online learning processes instances one by one and progressively learns from continuously arriving data to predict at the speed of the stream [49,73]. This approach offers responsiveness to changes in data streams while minimizing response times [73]. However, the major drawback is the limited view of the current stream, as a single instance may have a poor representation or be subject to noise.

On the other hand, chunk-based learning divides data into pieces or blocks, allowing for a better estimation of the current concept due to the size of the training set. However, the choice of block size is crucial. Blocks that are too small mean there aren't enough examples for effective model training, which can affect its ability to generalize properly from a limited set of examples. Conversely, blocks that are too large may lead to challenges in adapting to concept changes over time, posing drift management challenges [73].

It's worth noting that online algorithms are suited for applications dealing with real-time streaming data and facing time and memory constraints. Conversely, chunk-based and sliding window algorithms often use offline algorithms to learn from examples in each new block or window, which can result in poor predictive performance if each block or window contains only a single training example [3].

To address the shortcomings of these two approaches, hybrid approaches have been developed [3], leveraging online learning while retaining blocks of data to extract statistics and relevant information about the data stream. These data are then used for additional periodic updates to the model, providing an effective combination of the advantages of online learning and block-based methods.

5.2 Imbalanced Data

The issue of imbalanced data in supervised learning emerged over two decades ago [48] and remains a current problem. The concept of imbalance is commonly encountered in classification problems, where there are more instances of one class (majority) than another class (minority) [44]. For regression problems, imbalance is manifested by a significant difference in the number of instances in two value intervals.

To address data imbalance, several strategies have been proposed in the literature [3,44] (Fig. 4): 1) cost-sensitive strategy, 2) under/over-sampling, and 3) ensemble learning.

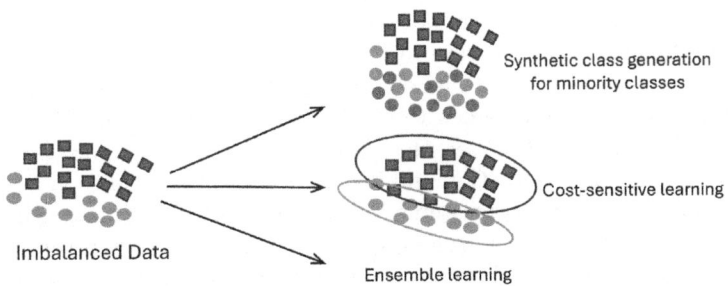

Fig. 4. Imbalanced data methods

Cost-sensitive strategy entails assigning higher weights to certain data instances in the loss function calculation. For example, when data is rare, it may be underrepresented in the learning process, leading to less accurate predictions for these instances. To address this underrepresentation and emphasize rare instances, the weights of prediction errors associated with these instances should be increased in the loss function calculation. This allows the model to predict these rare instances correctly, thereby reducing the bias introduced by underrepresentation.

Under/over-sampling of data involves either adding synthetic instances of minority classes in the classification problem or removing instances from a majority class to obtain a balanced dataset.

Ensemble learning also represents a solution to address issues of conceptual drift and imbalance. It involves combining multiple algorithms to enhance the learning model's performance, achieving a higher level of accuracy than would be achieved if any of these algorithms were used separately. Ensemble learning is often used with cost-sensitive learning or resampling strategies.

Classifying imbalanced data in data streams is a prevalent problem in the literature, and solutions to address it have focused on class resampling methods, adaptation algorithms, or ensemble methods [3]. Imbalanced data regression from streaming data is also subject to imbalance [5].

For the problem of analyzing imbalanced and non-stationary data streams, the literature offers a wide range of works on classification but fewer works for the regression task. One issue is related to adapting the model according to the imbalance ratio. Indeed, if the ratio is static, resampling methods and modifying model training can yield promising results [3].

5.3 Trustworthiness

An important step in the data analytics process is evaluating the reliability of generated predictions. Indeed, a trained model can produce erroneous predictions, especially when faced with situations it has not encountered before [20]. Therefore, it is crucial to quantify the level of trust we can have in the predictions produced by the models. This allows us to make decisions based on the associated confidence level of each prediction.

Trustworthiness or confidence in predictions lacks a universal definition. According to [53], it is related to its consistency with the training data, meaning the prediction is trustworthy when the training data exhibits similar behavior to that of the new instance. However, the work described in [20] makes a distinction between confidence in a prediction and its correctness. Confidence in a prediction is established based on its consistency with the training data, regardless of the ground truth, while correctness refers to the intrinsic precision of the prediction, which can only be assessed by comparing the prediction to the ground truth when available.

In the literature, there are works focusing on estimating uncertainty [56], interpretability [37], explainability [66] and trustworthiness [20,43,53]. These research domains are closely related and enable the assessment of the reliability of predictions generated by models. Interpretability [37,70] aims to explain and make a complex process or decision understandable for humans, as defined by [37], as well as uncertainty. These concepts can be interconnected, as better interpretability can help reinforce trust by reducing uncertainty related to decisions based on complex systems. Some studies [104] have examined individuals' ability to understand the relationship between the performance of a machine learning model on unseen data and its expected performance after deployment, as well as the issue of trust in decisions made by machine learning models. Other studies [43, 87] have focused on quantifying trust in predictions using quantitative measures of the quality of interpretability methods [87] or on specifying characteristics in deep neural networks [43]. These studies align with recent research on interpretable machine learning, considering different dimensions such as a model's internal components and predictions.

6 Conclusion and Future Directions

In this paper, we provide a comprehensive review of the role and impact of machine learning methods on the various stages of the data management process. We pay particular attention to the entire pipeline, including data preparation, integration, and analytics, with the goal of addressing challenges related to data quality, heterogeneity, non-stationarity, and imbalance.

Despite significant advancements in the field and the application of machine learning to address these challenges, several research issues remain. While our list of challenges is not exhaustive, we aim to outline future research directions for each phase.

6.1 Data Preparation

The quality of applications based on machine learning (ML) depends on the quality of the training data. Data cleaning has become crucial for building high-quality ML models [68]. Both the Big Data and ML communities are working on issues related to dirty data but from two different perspectives. The ML community focuses on understanding the impact of noise on ML models without performing data cleaning. Consequently, some ML algorithms have been developed to be robust against noise in certain distributions, such as decision trees, using regularization to improve robustness, or using the bagging model to reduce performance variability caused by noisy data [58]. On the other hand, the Data community has focused on understanding the fundamental process of data cleaning without considering its impact on ML models. We think that providing preparation tools "as services" for the community can help to improve ML processing pipelines. Moreover, with the advent of Data-Centric AI (DCAI) [106], the focus will be on developing more automated and efficient data preparation processes.

6.2 Data Integration

Multimodal Data Fusion: A modality refers to how we perceive and express natural phenomena through different data sources [69]. Multimodality refers to situations where multiple modalities are involved simultaneously. Multimodal learning aims to develop models and techniques capable of processing and exploiting heterogeneous data from different modalities (e.g., audio, image, text, video) to solve tasks such as classification. It requires the fusion and integration of these multiple data sources, taking into account:

- Their heterogeneity: Data often exhibit diverse characteristics, structures, and representations.
- Their interconnection: Relationships or similarities may exist between the data. For example, an image and its textual description may be connected because they describe the same phenomenon (e.g., a scene).
- Their interaction: When multiple modalities are integrated for a specific task or analysis, they interact in different ways, thus generating new information. This means that their combined use can lead to improved results compared to analyzing each modality in isolation.

Multimodal data fusion [40] aims to resolve conflicts between different data sources by integrating data with different distributions, sources, and types into a single space where different modalities (e.g., inter-modalities and cross-modalities) can be uniformly represented. This allows for richer information than that obtained from a single modality by leveraging information from different sources.

Future work will involve studying the interconnection and interactions between different modalities [69]. The main objective is to understand how to identify and analyze the links and exchanges of information between elements from different modalities.

6.3 Data Analytics

Uncertainty: Uncertainty is defined as a situation in which information may be unknown or imperfect. Managing uncertainty is not a recent problem. In the field of relational databases, uncertainty can exist at various levels such as data, schema, mapping of different schema instances, or even in user queries [84]. Several measures for uncertain data, such as uncertainty density or adaptations of precision and recall measures to evaluate the uncertainty of responses have been proposed in the literature [55].

With the advent of big data, uncertainty can arise from data collection, concept variance, and multimodality [47]. Its treatment must span multiple stages of the management process: from collection to machine learning and analysis.

Machine learning is naturally impacted by uncertainty as it involves extracting data to use them for prediction purposes. However, traditional machine learning algorithms do not necessarily consider uncertainty in data and predictions.

Uncertainty in machine learning can be quantified at the data level, known as aleatoric uncertainty, or at the model level, known as epistemic uncertainty [19,50]. Aleatoric (statistical) uncertainty is related to data and is irreducible due to the natural variability of random phenomena. Epistemic uncertainty refers to model uncertainty caused by a lack of knowledge about the training data, which can be reduced based on additional information.

Several methods for quantifying uncertainties have been proposed in the literature, especially in deep learning [25,41]: ensemble methods, Bayesian neural networks, evidential approaches, and Concrete Dropout. For all these methods, epistemic uncertainty is estimated by evaluating a set of trained models, although the approach to sampling these models differs between methods. Current deep learning uncertainty quantification methods do not allow for determining the possible causes of uncertainty. It is therefore crucial to understand how these methods operate and capture uncertainty to identify ways to improve, detect, and characterize it [41].

Context: Integrating context is crucial for more effective data learning process. According to Abowd et al. [1], context is defined as "any information that can be used to describe or characterize the situation of an entity". It encompasses aspects such as time, location, environment, relationships, etc. An entity can be an object, a system, a person, or any other element subject to particular analysis that may be considered relevant in an interaction between a user and an application. Considering context, especially in Large Language Models (LLM) is a broad research area that poses various challenges [110]. These challenges include relevant context selection, modeling, discovery as well as managing bias.

Human in the Loop: Human in the loop (HITL) refers to a model or computer system in which human intervention is required to complete a task or make decisions. This means that automation is present but supplemented by human involvement. The goal is often to leverage the benefits of artificial intelligence (AI) and automation while maintaining human supervision for reasons of safety, quality, or complexity.

Thus, we beleive that exploring the role of humans within the data management process represents an interesting challenge which aims to interactively guide the different phases of the process. For example, in data preparation, integrating human expertise can be important, particularly for correcting results generated by models [74]. Subsequently, the goal is to incorporate human expertise into the learning process since it plays a central role in decision-making. This could be, for instance, in the context of deep learning-based data fusion approaches, where the idea is to gradually merge modalities based on their similarity, guided by prior knowledge [95]. This raises major challenges, particularly concerning the formalization and integration of human knowledge within learning models and the selection of the most appropriate models [74,86].

The literature offers various approaches to address these challenges, including active learning for data labeling, automated learning, and interactive learning, as described in recent works [74].

References

1. Abowd, G.D., Dey, A.K., Brown, P.J., Davies, N., Smith, M., Steggles, P.: Towards a better understanding of context and context-awareness. In: Gellersen, H.W.

(ed.) Handheld and Ubiquitous Computing, pp. 304–307. Springer, Berlin Heidelberg, Berlin, Heidelberg (1999). https://doi.org/10.1007/3-540-48157-5_29
2. Ackerman, R., Gal, A., Sagi, T., Shraga, R.: A cognitive model of human bias in matching. In: Nayak, A.C., Sharma, A. (eds.) PRICAI 2019. LNCS (LNAI), vol. 11670, pp. 632–646. Springer, Cham (2019). https://doi.org/10.1007/978-3-030-29908-8_50
3. Aguiar, G., Krawczyk, B., Cano, A.: A survey on learning from imbalanced data streams: taxonomy, challenges, empirical study, and reproducible experimental framework. CoRR abs/ arxiv:2204.03719 (2022)
4. Al-Sai, Z.A., et al.: Explore big data analytics applications and opportunities: A review. Big Data Cogn. Comput. **6**(4), 157 (2022)
5. Aminian, E., Ribeiro, R.P., Gama, J.: Chebyshev approaches for imbalanced data streams regression models. Data Min. Knowl. Discov. **35**(6), 2389–2466 (2021)
6. Anava, O., Hazan, E., Zeevi, A.: Online time series prediction with missing data. In: Bach, F.R., Blei, D.M. (eds.) Proceedings of the 32nd International Conference on Machine Learning, ICML 2015, Lille, France, 6-11 July 2015. JMLR Workshop and Conference Proceedings, vol. 37, pp. 2191–2199. JMLR.org (2015)
7. Bahri, M., Bifet, A., Gama, J., Gomes, H.M., Maniu, S.: Data stream analysis: foundations, major tasks and tools. WIREs Data Mining Knowl. Discov. **11**(3) (2021)
8. Balzano, L., Chi, Y., Lu, Y.M.: Streaming PCA and subspace tracking: the missing data case. Proc. IEEE **106**(8), 1293–1310 (2018)
9. Barlaug, N., Gulla, J.A.: Neural networks for entity matching: a survey. ACM Trans. Knowl. Discov. Data **15**(3), 52:1–52:37 (2021)
10. Batini, C., Cappiello, C., Francalanci, C., Maurino, A.: Methodologies for data quality assessment and improvement. ACM Comput. Surv. **41**(3), 16:1–16:52 (2009)
11. Batko, K.M., Slezak, A.: The use of big data analytics in healthcare. J. Big Data **9**(1), 3 (2022)
12. Bellahsene, Z., Bonifati, A., Rahm, E. (eds.): Schema Matching and Mapping. Springer, Data-Centric Systems and Applications (2011)
13. Bellare, K., Iyengar, S., Parameswaran, A.G., Rastogi, V.: Active sampling for entity matching. In: Yang, Q., Agarwal, D., Pei, J. (eds.) The 18th ACM SIGKDD International Conference on Knowledge Discovery and Data Mining, KDD 2012, Beijing, China, 12-16 August 2012, pp. 1131–1139. ACM (2012)
14. Bernstein, P.A., Madhavan, J., Rahm, E.: Generic schema matching, ten years later. Proc. VLDB Endow. **4**(11), 695–701 (2011)
15. Berti-Équille, L.: Learn2clean: Optimizing the sequence of tasks for web data preparation. In: Liu, L., White, R.W., Mantrach, A., Silvestri, F., McAuley, J.J., Baeza-Yates, R., Zia, L. (eds.) The World Wide Web Conference, WWW 2019, San Francisco, CA, USA, May 2019, pp. 2580–2586. ACM (2019)
16. Berti-Équille, L., Bonifati, A., Milo, T.: Machine learning to data management: a round trip. In: 34th IEEE International Conference on Data Engineering, ICDE 2018, Paris, France, 16-19 April 2018, pp. 1735–1738. IEEE Computer Society (2018)
17. Berti-Équille, L., Dasu, T., Srivastava, D.: Discovery of complex glitch patterns: A novel approach to quantitative data cleaning. In: Abiteboul, S., Böhm, K., Koch, C., Tan, K. (eds.) Proceedings of the 27th International Conference on Data Engineering, ICDE 2011, 11-16 April 2011, Hannover, Germany. pp. 733–744. IEEE Computer Society (2011)

18. Beyer, C., Büttner, M., Spiliopoulou, M.: Challenges for active feature acquisition and imputation on data streams. In: Bunse, M., Hammer, B., Krempl, G., Lemaire, V., Tharwat, A., Saadallah, A. (eds.) Proceedings of the Workshop on Interactive Adaptive Learning co-located with European Conference on Machine Learning and Principles and Practice of Knowledge Discovery in Databases (ECML-PKDD 2023), Torino, Italy, September 22nd, 2023. CEUR Workshop Proceedings, vol. 3470, pp. 9–13. CEUR-WS.org (2023)
19. Bhatt, U., Antorán, J., Zhang, Y., Liao, Q.V., Sattigeri, P., Fogliato, R., Melançon, G.G., Krishnan, R., Stanley, J., et al.: Uncertainty as a form of transparency: Measuring, communicating, and using uncertainty. In: Fourcade, M., Kuipers, B., Lazar, S., Mulligan, D.K. (eds.) AIES 2021: AAAI/ACM Conference on AI, Ethics, and Society, Virtual Event, USA, 19-21 May 2021, pp. 401–413. ACM (2021)
20. de Bie, K., Lucic, A., Haned, H.: To trust or not to trust a regressor: Estimating and explaining trustworthiness of regression predictions. CoRR abs/arxiv:2104.06982 (2021)
21. Boniol, P., Paparrizos, J., Palpanas, T.: New trends in time series anomaly detection. In: Stoyanovich, J., et al.: (eds.) Proceedings 26th International Conference on Extending Database Technology, EDBT 2023, Ioannina, Greece, 28-31 March 2023, pp. 847–850. OpenProceedings.org (2023)
22. Braei, M., Wagner, S.: Anomaly detection in univariate time-series: A survey on the state-of-the-art. arXiv preprint arXiv:2004.00433 (2020)
23. Brunner, U., Stockinger, K.: Entity matching on unstructured data: an active learning approach. In: 6th Swiss Conference on Data Science, SDS 2019, Bern, Switzerland, 14 June 2019, pp. 97–102. IEEE (2019)
24. Brunner, U., Stockinger, K.: Entity matching with transformer architectures - A step forward in data integration. In: Bonifati, A., et al. (eds.) Proceedings of the 23rd International Conference on Extending Database Technology, EDBT 2020, Copenhagen, Denmark, 30 March - 02 April 2020. pp. 463–473. OpenProceedings.org (2020)
25. Caldeira, J., Nord, B.: Deeply uncertain: comparing methods of uncertainty quantification in deep learning algorithms. Mach. Learn. Sci. Technol. **2**(1), 15002 (2021)
26. Castanedo, F.: A review of data fusion techniques **2013**, 704504 (2013)
27. Chandola, V., Banerjee, A., Kumar, V.: Anomaly detection: a survey. ACM Comput. Surv. (CSUR) **41**(3), 1–58 (2009)
28. Christen, V., Christen, P., Rahm, E.: Informativeness-based active learning for entity resolution. In: Cellier, P., Driessens, K. (eds.) ECML PKDD 2019. CCIS, vol. 1168, pp. 125–141. Springer, Cham (2020). https://doi.org/10.1007/978-3-030-43887-6_11
29. Christophides, V., Efthymiou, V., Palpanas, T., Papadakis, G., Stefanidis, K.: End-to-end entity resolution for big data: A survey. CoRR abs/arXiv: 1905.06397 (2019)
30. Christophides, V., Efthymiou, V., Palpanas, T., Papadakis, G., Stefanidis, K.: An overview of end-to-end entity resolution for big data. ACM Comput. Surv. **53**(6), 127:1–127:42 (2021)
31. Chu, X., Ilyas, I.F.: Qualitative data cleaning. Proc. VLDB Endow. **9**(13), 1605–1608 (2016)
32. Chu, X., Ilyas, I.F., Krishnan, S., Wang, J.: Data cleaning: overview and emerging challenges. In: Özcan, F., Koutrika, G., Madden, S. (eds.) Proceedings of the 2016

International Conference on Management of Data, SIGMOD Conference 2016, San Francisco, CA, USA, 26 June - 01 July 2016. pp. 2201–2206. ACM (2016)
33. Cohen, W.W., Ravikumar, P., Fienberg, S.E.: A comparison of string distance metrics for name-matching tasks. In: Kambhampati, S., Knoblock, C.A. (eds.) Proceedings of IJCAI-03 Workshop on Information Integration on the Web (IIWeb-03), 9-10 August 2003, Acapulco, Mexico, pp. 73–78 (2003)
34. David, J., Euzenat, J., Šváb-Zamazal, O.: Ontology similarity in the alignment space. In: Patel-Schneider, P.F., et al. (eds.) ISWC 2010. LNCS, vol. 6496, pp. 129–144. Springer, Heidelberg (2010). https://doi.org/10.1007/978-3-642-17746-0_9
35. Dong, L., Rekatsinas, T.: Data integration and machine learning: a natural synergy. Proc. VLDB Endow. **11**(12), 2094–2097 (2018)
36. Dong, X.L., Naumann, F.: Data fusion - resolving data conflicts for integration. Proc. VLDB Endow. **2**(2), 1654–1655 (2009)
37. Du, M., Liu, N., Hu, X.: Techniques for interpretable machine learning. Commun. ACM **63**(1), 68–77 (2020)
38. Ester, M., Kriegel, H.P., Sander, J., Xu, X., et al.: A density-based algorithm for discovering clusters in large spatial databases with noise. In: kdd, vol. 96, pp. 226–231 (1996)
39. Fekade, B., Maksymyuk, T., Kyryk, M., Jo, M.: Probabilistic recovery of incomplete sensed data in IoT. IEEE Internet Things J. (2018)
40. Gao, J., Li, P., Chen, Z., Zhang, J.: A survey on deep learning for multimodal data fusion. Neural Comput. **32**(5), 829–864 (2020)
41. Gawlikowski, J., et al.: A survey of uncertainty in deep neural networks. CoRR abs/ arXiv: 2107.03342 (2021)
42. Ge, C., Wang, P., Chen, L., Liu, X., Zheng, B., Gao, Y.: Collaborer: a self-supervised entity resolution framework using multi-features collaboration. CoRR abs/ arXiv: 2108.08090 (2021)
43. Ghobrial, A., Hond, D., Asgari, H., Eder, K.: A trustworthiness score to evaluate DNN predictions. In: IEEE International Conference on Artificial Intelligence Testing, AITest 2023, Athens, Greece, 17-20 July 2023, pp. 9–16. IEEE (2023)
44. Gomes, H.M., Read, J., Bifet, A., Barddal, J.P., Gama, J.: Machine learning for streaming data: state of the art, challenges, and opportunities. ACM SIGKDD Explorations Newsl **21**(2), 6–22 (2019)
45. Goswami, S., Kumar, A.: Survey of deep-learning techniques in big-data analytics. Wirel. Pers. Commun. **126**(2), 1321–1343 (2022)
46. Hameed, M., Naumann, F.: Data preparation: a survey of commercial tools. SIGMOD Rec. **49**(3), 18–29 (2020)
47. Hariri, R.H., Fredericks, E.M., Bowers, K.M.: Uncertainty in big data analytics: survey, opportunities, and challenges. J. Big Data **6**, 44 (2019)
48. He, H., Garcia, E.A.: Learning from imbalanced data. IEEE Trans. Knowl. Data Eng. **21**(9), 1263–1284 (2009)
49. Hoi, S.C., Sahoo, D., Lu, J., Zhao, P.: Online learning: A comprehensive survey. arXiv preprint arXiv:1802.02871 (2018)
50. Hüllermeier, E., Waegeman, W.: Aleatoric and epistemic uncertainty in machine learning: an introduction to concepts and methods. Mach. Learn. **110**(3), 457–506 (2021)
51. Ilyas, I.F., Rekatsinas, T.: Machine learning and data cleaning: which serves the other? ACM J. Data Inform. Quality (JDIQ) (2022)
52. Jain, A., Sarawagi, S., Sen, P.: Deep indexed active learning for matching heterogeneous entity representations. Proc. VLDB Endow. **15**(1), 31–45 (2021)

53. Jiang, H., Kim, B., Guan, M.Y., Gupta, M.R.: To trust or not to trust A classifier. In: Bengio, S., Wallach, H.M., Larochelle, H., Grauman, K., Cesa-Bianchi, N., Garnett, R. (eds.) Advances in Neural Information Processing Systems 31: Annual Conference on Neural Information Processing Systems 2018, NeurIPS 2018, 3-8 December 2018, Montréal, Canada, pp. 5546–5557 (2018)
54. Kaisler, S.H., Espinosa, J.A., Money, W.H., Armour, F.: Big data and analytics: issues and challenges for the past and next ten years. In: Bui, T.X. (ed.) 56th Hawaii International Conference on System Sciences, HICSS 2023, Maui, Hawaii, USA, 3-6 January 2023. pp. 805–814. ScholarSpace (2023)
55. de Keijzer, A., van Keulen, M.: Quality measures in uncertain data management. In: Prade, H., Subrahmanian, V.S. (eds.) SUM 2007. LNCS (LNAI), vol. 4772, pp. 104–115. Springer, Heidelberg (2007). https://doi.org/10.1007/978-3-540-75410-7_8
56. Kendall, A., Gal, Y.: What uncertainties do we need in bayesian deep learning for computer vision? In: Guyon, I., von Luxburg, U., Bengio, S., Wallach, H.M., Fergus, R., Vishwanathan, S.V.N., Garnett, R. (eds.) Advances in Neural Information Processing Systems 30: Annual Conference on Neural Information Processing Systems 2017, 4-9 December 2017, Long Beach, CA, USA, pp. 5574–5584 (2017)
57. Khayati, M., Arous, I., Tymchenko, Z., Cudré-Mauroux, P.: ORBITS: online recovery of missing values in multiple time series streams. Proc. VLDB Endow. **14**(3), 294–306 (2020)
58. Khoshgoftaar, T.M., Hulse, J.V., Napolitano, A.: Comparing boosting and bagging techniques with noisy and imbalanced data. IEEE Trans. Syst. Man Cybern. Part A **41**(3), 552–568 (2011)
59. Konda, P., et al.: Magellan: toward building entity matching management systems. Proceedings VLDB Endowment **9**(12), 1197–1208 (2016)
60. Konstantinou, N., Paton, N.W.: Feedback driven improvement of data preparation pipelines. Inf. Syst. **92**, 101480 (2020)
61. Koutras, C., Siachamis, G., Ionescu, A., Psarakis, K., Brons, J., Fragkoulis, M., Lofi, C., Bonifati, A., Katsifodimos, A.: Valentine: Evaluating matching techniques for dataset discovery. In: 37th IEEE International Conference on Data Engineering, ICDE 2021, Chania, Greece, 19-22 April 2021, pp. 468–479. IEEE (2021)
62. Krishnan, S., Wu, E.: Alphaclean: Automatic generation of data cleaning pipelines. CoRR abs/ arxiv: 1904.11827 (2019), http://arxiv.org/abs/1904.11827
63. Kumar, A., Boehm, M., Yang, J.: Data management in machine learning: challenges, techniques, and systems. In: Salihoglu, S., Zhou, W., Chirkova, R., Yang, J., Suciu, D. (eds.) Proceedings of the 2017 ACM International Conference on Management of Data, SIGMOD Conference 2017, Chicago, IL, USA, 14-19 May 2017, pp. 1717–1722. ACM (2017)
64. Labrinidis, A., Jagadish, H.V.: Challenges and opportunities with big data. Proc. VLDB Endow. **5**(12), 2032–2033 (2012)
65. L'Heureux, A., Grolinger, K., ElYamany, H.F., Capretz, M.A.M.: Machine learning with big data: challenges and approaches. IEEE Access **5**, 7776–7797 (2017)
66. Li, D., Liu, Y., Huang, J., Wang, Z.: A trustworthy view on explainable artificial intelligence method evaluation. Computer **56**(4), 50–60 (2023)
67. Li, G., Wang, J., Zheng, Y., Franklin, M.J.: Crowdsourced data management: a survey. In: 33rd IEEE International Conference on Data Engineering, ICDE 2017, San Diego, CA, USA, 19-22 April 2017, pp. 39–40. IEEE Computer Society (2017)

68. Li, P., Rao, X., Blase, J., Zhang, Y., Chu, X., Zhang, C.: Cleanml: a study for evaluating the impact of data cleaning on ML classification tasks. In: 37th IEEE International Conference on Data Engineering, ICDE 2021, Chania, Greece, 19-22 April 2021, pp. 13–24. IEEE (2021)
69. Liang, P.P., Zadeh, A., Morency, L.: Foundations and recent trends in multimodal machine learning: Principles, challenges, and open questions. CoRR **abs/2209.03430** (2022). https://doi.org/10.48550/ARXIV.2209.03430
70. Lipton, Z.C.: The mythos of model interpretability. Commun. ACM **61**(10), 36–43 (2018)
71. Mahdavinejad, M.S., Rezvan, M., Barekatain, M., Adibi, P., Barnaghi, P.M., Sheth, A.P.: Machine learning for internet of things data analysis: A survey. CoRR abs/ arXiv:1802.06305 (2018)
72. Marjani, M., et al.: Big iot data analytics: architecture, opportunities, and open research challenges. IEEE Access **5**, 5247–5261 (2017)
73. Minku, L.L.: Transfer Learning in Non-stationary Environments, pp. 13–37. Springer International Publishing, Cham (2019). https://doi.org/10.1007/978-3-319-89803-2_2
74. Mosqueira-Rey, E., Hernández-Pereira, E., Alonso-Ríos, D., Bobes-Bascarán, J., Fernández-Leal, Á.: Human-in-the-loop machine learning: a state of the art. Artif. Intell. Rev. **56**(4), 3005–3054 (2023)
75. Nti, I.K., Quarcoo, J.A., Aning, J., Fosu, G.K.: A mini-review of machine learning in big data analytics: applications, challenges, and prospects. Big Data Min. Anal. **5**(2), 81–97 (2022)
76. Papadakis, G., Ioannou, E., Palpanas, T.: Entity resolution: Past, present and yet-to-come. In: EDBT (2020)
77. Papadakis, G., Ioannou, E., Thanos, E., Palpanas, T.: The Four Generations of Entity Resolution. Morgan & Claypool Publishers, Synthesis Lectures on Data Management (2021)
78. Papadakis, G., Tsekouras, L., Thanos, E., Giannakopoulos, G., Palpanas, T., Koubarakis, M.: Domain-and structure-agnostic end-to-end entity resolution with jedai. ACM SIGMOD Rec. **48**(4), 30–36 (2020)
79. Peng, T., Sellami, S., Boucelma, O.: Iot data imputation with incremental multiple linear regression. Open J. Internet Things **5**(1), 69–79 (2019)
80. Rahm, E., Bernstein, P.A.: A survey of approaches to automatic schema matching. VLDB J. **10**(4), 334–350 (2001)
81. Rahm, E., Do, H.H.: Data cleaning: problems and current approaches. IEEE Data Eng. Bull. **23**(4), 3–13 (2000)
82. Rattenbury, T., Hellerstein, J.M., Heer, J., Kandel, S., Carreras, C.: Principles of data wrangling: Practical techniques for data preparation. "O'Reilly Media, Inc." (2017)
83. Ré, C., Agrawal, D., Balazinska, M., Cafarella, M.J., Jordan, M.I., Kraska, T., Ramakrishnan, R.: Machine learning and databases: the sound of things to come or a cacophony of hype? In: Sellis, T.K., Davidson, S.B., Ives, Z.G. (eds.) Proceedings of the 2015 ACM SIGMOD International Conference on Management of Data, Melbourne, Victoria, Australia, 31 May - 4 June 2015. pp. 283–284. ACM (2015)
84. Ré, C., Suciu, D.: Management of data with uncertainties. In: Silva, M.J., et al. (eds.) Proceedings of the Sixteenth ACM Conference on Information and Knowledge Management, CIKM 2007, Lisbon, Portugal, 6-10 November 2007, pp. 3–8. ACM (2007)

85. Rekatsinas, T., Chu, X., Ilyas, I.F., Ré, C.: Holoclean: holistic data repairs with probabilistic inference. Proc. VLDB Endow. **10**(11), 1190–1201 (2017)
86. von Rueden, L., et al.: Informed machine learning - a taxonomy and survey of integrating prior knowledge into learning systems. IEEE Trans. Knowl. Data Eng. **35**(01), 614–633 (2023). https://doi.org/10.1109/TKDE.2021.3079836
87. Schmidt, P., Bießmann, F.: Quantifying interpretability and trust in machine learning systems. CoRR abs/ arXiv: 1901.08558 (2019)
88. Shraga, R.: (artificial) mind over matter: Humans in and humans out in matching. In: Abedjan, Z., Hose, K. (eds.) Proceedings of the VLDB 2020 PhD Workshop co-located with the 46th International Conference on Very Large Databases (VLDB 2020), ONLINE, 31 August - 4 September 020. CEUR Workshop Proceedings, vol. 2652. CEUR-WS.org (2020)
89. Shvaiko, P., Euzenat, J.: A survey of schema-based matching approaches. J. Data Semant., 146–171 (2005). https://doi.org/10.1007/11603412_5
90. Shvaiko, P., Euzenat, J.: Ontology matching: State of the art and future challenges. IEEE Trans. Knowl. Data Eng. **25**(1), 158–176 (2013)
91. Siddiqa, A., et al.: A survey of big data management: taxonomy and state-of-the-art. J. Netw. Comput. Appl. **71**, 151–166 (2016)
92. bibitemch1SidiPAJIM12 Sidi, F., Panah, P.H.S., Affendey, L.S., Jabar, M.A., Ibrahim, H., Mustapha, A.: Data quality: a survey of data quality dimensions. In: Mahmod, R., et al. (eds.) 2012 International Conference on Information Retrieval & Knowledge Management, Kuala Lumpur, Malaysia, 13-15 March 2012, pp. 300–304. IEEE (2012)
93. Somasundaram, R., Nedunchezhian, R.: Evaluation of three simple imputation methods for enhancing preprocessing of data with missing values. Inter. J. Comput. Appli. **21**(10) (2011)
94. Song, S., Zhang, A.: Iot data quality. In: d'Aquin, M., Dietze, S., Hauff, C., Curry, E., Cudré-Mauroux, P. (eds.) CIKM 2020: The 29th ACM International Conference on Information and Knowledge Management, Virtual Event, Ireland, 19-23 October 2020, pp. 3517–3518. ACM (2020)
95. Stahlschmidt, S., Ulfenborg, B., Synnergren, J.: Multimodal deep learning for biomedical data fusion: a review. Briefings Bioinform. **23** (01 2022)
96. Thirumuruganathan, S., Parambath, S.A.P., Ouzzani, M., Tang, N., Joty, S.R.: Reuse and adaptation for entity resolution through transfer learning. CoRR abs/ arXiv: 1809.11084 (2018)
97. Thudumu, S., Branch, P., Jin, J., Singh, J.: A comprehensive survey of anomaly detection techniques for high dimensional big data. J. Big Data **7**, 1–30 (2020)
98. Tsai, C., Lai, C., Chao, H., Vasilakos, A.V.: Big data analytics: a survey. J. Big Data **2**, 21 (2015)
99. Turabieh, H., Salem, A.A., Abu-El-Rub, N.: Dynamic l-rnn recovery of missing data in iomt applications. Futur. Gener. Comput. Syst. **89**, 575–583 (2018)
100. Vassiliadis, P.: A survey of extract-transform-load technology. Inter. J. Data Warehousing Mining (IJDWM) **5**(3), 1–27 (2009)
101. Wang, J., Kraska, T., Franklin, M.J., Feng, J.: Crowder: crowdsourcing entity resolution. Proc. VLDB Endow. **5**(11), 1483–1494 (2012)
102. Wang, W., Zhang, M., Chen, G., Jagadish, H.V., Ooi, B.C., Tan, K.: Database meets deep learning: challenges and opportunities. SIGMOD Rec. **45**(2), 17–22 (2016)
103. Wellenzohn, K., Böhlen, M.H., Dignös, A., Gamper, J., Mitterer, H.: Continuous imputation of missing values in streams of pattern-determining time series.

In: Markl, V., Orlando, S., Mitschang, B., Andritsos, P., Sattler, K., Breß, S. (eds.) Proceedings of the 20th International Conference on Extending Database Technology, EDBT 2017, Venice, Italy, 21-24 March 2017, pp. 330–341. OpenProceedings.org (2017)
104. Yin, M., Vaughan, J.W., Wallach, H.M.: Understanding the effect of accuracy on trust in machine learning models. In: Brewster, S.A., Fitzpatrick, G., Cox, A.L., Kostakos, V. (eds.) Proceedings of the 2019 CHI Conference on Human Factors in Computing Systems, CHI 2019, Glasgow, Scotland, UK,04-09 May 2019, p. 279. ACM (2019)
105. Yoon, J., Zame, W.R., van der Schaar, M.: Estimating missing data in temporal data streams using multi-directional recurrent neural networks. IEEE Trans. Biomed. Eng. **66**(5), 1477–1490 (2019)
106. Zha, D., Bhat, Z.P., Lai, K., Yang, F., Hu, X.: Data-centric AI: perspectives and challenges. In: Shekhar, S., Zhou, Z., Chiang, Y., Stiglic, G. (eds.) Proceedings of the 2023 SIAM International Conference on Data Mining, SDM 2023, Minneapolis-St. Paul Twin Cities, MN, USA, 27-29 April 2023, pp. 945–948. SIAM (2023)
107. Zhang, C.J., Chen, L., Jagadish, H.V., Zhang, M., Tong, Y.: Reducing uncertainty of schema matching via crowdsourcing with accuracy rates. IEEE Trans. Knowl. Data Eng. **32**(1), 135–151 (2020)
108. Zhang, Y., Floratou, A., Cahoon, J., Krishnan, S., Müller, A.C., Banda, D., Psallidas, F., Patel, J.M.: Schema matching using pre-trained language models. In: 39th IEEE International Conference on Data Engineering, ICDE 2023, Anaheim, CA, USA, 3-7 April 2023. pp. 1558–1571. IEEE (2023)
109. Zhao, Y., Landgrebe, E., Shekhtman, E., Udell, M.: Online missing value imputation and change point detection with the gaussian copula. In: Thirty-Sixth AAAI Conference on Artificial Intelligence, AAAI 2022, Thirty-Fourth Conference on Innovative Applications of Artificial Intelligence, IAAI 2022, The Twelveth Symposium on Educational Advances in Artificial Intelligence, EAAI 2022 Virtual Event, 22 February - 1 March 2022. pp. 9199–9207. AAAI Press (2022)
110. Zhu, Y., Moniz, J.R.A., Bhargava, S., Lu, J., Piraviperumal, D., Li, S., Zhang, Y., Yu, H., Tseng, B.: Can large language models understand context? In: Graham, Y., Purver, M. (eds.) Findings of the Association for Computational Linguistics: EACL 2024, St. Julian's, Malta, 17-22 March 2024, pp. 2004–2018. Association for Computational Linguistics (2024)

Exploring Reciprocal Exchanges and Trust-Based Authorizations: A Feasibility Demonstration with Location-Based Services

Gabriela Suntaxi[1(✉)], Aboubakr Achraf El Ghazi[2], and Klemens Böhm[3]

[1] Department of Informatics and Computer Science, National Polytechnic School, Quito, Ecuador
gabriela.suntaxi@epn.edu.ec
[2] German Center for Aviation and Space Flight, Stuttgart, Baden-Württemberg, Germany
[3] Karlsruhe Institute of Technology, Am Fasanengarten 5, Karlsruhe, Germany

Abstract. There is a large body of evidence in fields like psychology and sociology indicating that reciprocity is a powerful determinant of human behavior. However, none of the existing access control models captures this reciprocity phenomenon. In this paper, we introduce a new decision-type, which we call *reciprocal*, to by which users grant access to their resources only to those other users who allow them reciprocal access. We define the syntax and semantics of *reciprocal* authorizations and show how to include this new decision-type in the Attribute-Based Access Control model. We use location-based services as an example to deploy *reciprocal* authorizations; we propose two approaches to integrate them into these services and analyze their soundness and complexity. Next, we prove the soundness and analyze the complexity of both approaches. We also study how the ratio of *reciprocal* to *allow* and to *deny* authorizations affects the number of persons whose position a given person may read. These ratios may help in predicting whether users are willing to use *reciprocal* authorizations instead of *deny* or *allow*. Experiments confirm our complexity analyses and shed light on the performance of our approaches.

Keywords: access control models · reciprocity · reciprocal authorizations

1 Introduction

1.1 Motivation

Protecting information from unauthorized access is essential to guarantee data confidentiality. Access policies, delineating resource accessibility, are typically

---

Most part of this work was done while the first and second authors were at the Kalrsruhe Institute of Technology.

enforced through authorizations within specific access control models. Various models have been proposed; a prominent one is Attribute-based Access Control (ABAC) [1]. Traditional access policies are designed to fulfill the needs of an organization or the individual needs of the users of a system. However, social dynamics and human behavior play a significant role when users determine resource access. Reciprocity, a key trend in psychology, economics, and sociology, suggests that individuals often respond reciprocally to friendly actions. Studies [2,3] show that humans, despite inherent self-interest, often engage in cooperative behavior triggered by reciprocity This concept extends to access control, where individuals grant access to their resources to others who reciprocate the gesture. For example, in a bike-sharing system, users are more inclined to allow others to use their bikes if those users reciprocally permit them to use theirs.

Example 1. Consider a bike-sharing system where Anne and Bob own bikes, denoted as *bikeA* and *bikeB* respectively. Anne wishes to allow Bob to locate and use *bikeA* only if Bob grants her permission to locate and use *bikeB*. In conventional access control models, Anne must individually verify and set access privileges for each user, like Bob. Extending this scenario to include 10,000 individuals, each owning a bike, Anne aims to permit any of them, denoted as I, to access her bike, on the condition that I grants her access to their bike. However, repeating this process for each individual is impractical and sensitive information exposure is a concern. Moreover, continuous monitoring of access policies is required. With the *reciprocal* authorization feature, Anne simplifies this process by specifying just one authorization, significantly streamlining access management.

Existing access control models do not explicitly support reciprocity. Although there exist various models [4–6], they all consider only two decision-types, *allow* and *deny*. But for reciprocity, a new decision-type is needed. We call it *reciprocal* and authorizations that make use of it *reciprocal authorizations*.

Depending on the domain, access control models can protect different resources. Consider location-based services (LBS). Here, protecting the physical position of the users is crucial. They can be used to infer other personal information such as political affiliations, state of health, or personal preferences [7]. Users often are unwilling to share their position with all users in the system. But it is natural to share it with users willing to share theirs.

Reciprocal authorizations facilitate equitable resource exchange across various collaborative environments. They enable balanced sharing of information types on social media platforms, including personal data, images, and comments. Likewise, in academic circles, research groups can promote cooperation by exchanging resources like computational capabilities and datasets, contingent upon reciprocity. This concept extends to sensitive domains such as health records and browsing histories, where users may agree to reciprocally share information. Moreover, reciprocal authorizations enhance security by offering selective and transparent access control mechanisms. Selective access ensures resources are shared only with trusted individuals, reducing the risk of unauthorized access. Meanwhile, transparent access provides visibility into user access decisions and actions, empowering users to effectively monitor and control access.

This paper introduces the syntax and semantics of *reciprocal* authorizations, suitable for integration into any access control model. ABAC is specifically chosen for its relevance, showcasing how *reciprocal* authorizations can be conceptually integrated. Utilizing location-based services (LBSs) as an example, we illustrate the deployment of *reciprocal* authorizations, with a focus on k-nearest neighbor and range queries.

1.2 Challenges

Defining the syntax and semantics of *reciprocal* authorizations is subject to several challenges. First, the entire semantics of authorizations has to be redefined to incorporate a decision-type that supports reciprocity. For *reciprocal* authorizations, in particular, the semantics are not trivial. While reciprocity is observed in social network interactions like friending, extending this concept to access control is unprecedented, as it typically relies on *allow* and *deny* actions only.

A second challenge is the complexity of the computations required to give access to a resource. In models with only *allow* and *deny* authorizations, to determine who can access the resources of a user u, it is enough to know all authorizations that u has specified. With *reciprocal* authorizations, however, one also must know all authorizations assigned to u.

The main idea of *reciprocal* authorizations is to support a fair exchange of resources. The third challenge is to guarantee such a fairness principle, especially if the resources involved in the exchange have different degrees of sensitivity.

The fourth challenge lies in efficiently integrating services, such as LBS, with access policies while preserving existing implementations. Determining the outcome of a service with authorizations in place is non-trivial. Verification of authorizations and determining user access to required resources are necessary for data confidentiality. However, integrating reciprocal authorizations into services offers multiple approaches, making the processing of queries unclear.

Example 2. Consider a LBS provider (LBSP). A user has issued a location-dependent query that the LBSP now executes, knowing the user position. How can the LBSP take in authorizations to answer such queries? The result of such a query is the set of persons who fulfill the query constraint, and whose positions the querying user may see. Various integration designs are conceivable. One approach is to initially execute the query and subsequently filter the result based on authorizations. Alternatively, the LBSP could first compute the *view of the user*, i.e., the set of positions accessible to the querying user, and then execute the query on this view. Deciding which alternative is better is not trivial.

Lastly, the impact of the ratio of *reciprocal* to *allow* and *deny* authorizations on user views remains uncertain. This ratio is crucial in predicting whether users might prefer *reciprocal* authorizations over *deny* or *allow*. While it is expected that replacing *deny* authorizations with *reciprocal* ones could increase user views, the magnitude of this effect is unclear. Computing this based on the ratio of *reciprocal* to *deny* and *allow* authorizations presents challenges.

Example 3. Suppose given the percentages of *allow* and *deny* authorizations, the probability P_s that a randomly chosen user s can view the positions of 10% of the population is 0.4. Assume further that, if 10% of *deny* authorizations are replaced by *reciprocal*, P_s increases to 0.8. This suggests that users might consider using *reciprocal* authorizations instead of *deny*, as replacing only 10% of *deny* authorizations with *reciprocal* doubles P_s. However, if 90% replacement were needed, the opposite effect might occur.

1.3 Contributions

First, we introduce a new decision-type termed *reciprocal*, facilitating the modeling of reciprocal behavior. We define the syntax and semantics of *reciprocal* authorizations and illustrate their conceptual integration into ABAC. For deployment, we select Location-Based Services (LBSs) due to their clear semantics, contrasting with more complex scenarios like similarity queries for health records, where fairness in information exchange is challenging. For instance, Anne may not find it fair to open her health record if she has a stigmatizing disease in exchange for looking at the record of Bob, who is in perfect health. However, whether a disease is stigmatizing or not is subjective.

Second, we extend our basic model of *reciprocal* authorizations to accommodate scenarios with varying resource sensitivity. Assessing the sensitivity similarity between two resources remains a challenging task with no automated solution available. Inspired by trust models in the literature [8,9], our solution allows resource owners to assign sensitivity levels, which peers can subsequently evaluate. These evaluations contribute to computing a trust value for each user, empowering them to set minimum trust thresholds for resource exchange. We call this extension trust- based authorization.

Third, we integrate *reciprocal* authorizations into LBSs through two approaches: *filtering-querying* and *querying-filtering*. We prove the soundness of both approaches and analyze their complexity to ascertain their optimal conditions The analysis shows that there is no clear winner because the outcome depends on the query constraints, the dataset size, and the view size of the querying user. Knowing these parameters, our model can say which approach is better. Next, we analyze how the difference in the ratio of *reciprocal* to *deny* and *allow* authorizations affects the view size of a user. Finally, we conduct experiments to evaluate the impact of *reciprocal* authorizations on access decision performance, validate our complexity analysis, and assess the effectiveness of our approaches. Our experiments confirm the reliability of our complexity analysis in predicting algorithm behavior.

This paper extends the findings of a previous publication [10] by addressing issues related to unfair exchanges inherent in *reciprocal* authorizations (Sects. 4 and 5). This article also features two design alternatives for integrating LBSs with *reciprocal* authorizations, proves their soundness, and conducts time complexity analyses of them (Sects. 6 and 7). We also study how the ratio of *reciprocal*, *deny*, and *allow* authorizations affects the view of a user and evaluate our findings experimentally (Sect. 8). Finally, we conduct experiments to evaluate the impact of *reciprocal* authorizations on an access decision performance and the efficiency of the proposed approaches (Sect. 9).

2 Background and Notation

Access control models have been studied widely, and several models have been proposed [4,5]. In recent years, ABAC has gained considerable attention from business, academia and standard bodies. ABAC is more flexible for expressing access conditions than models such as role- based access control model. Therefore, it is adequate to support access conditions in LBS scenarios such as our bike- sharing scenario of Example 1. The following describes ABAC based on the terminology used by NIST [1].

2.1 ABAC Components

ABAC uses attributes associated with individuals and resources to specify authorizations that establish who can perform an operation, e.g., read or write, on a certain resource. We now formally present the main components of ABAC:

- **Attribute.** An attribute is a characteristic that defines specific aspects of a subject or a resource. $Attr_{Entity}$ is the set of all attributes of an entity. The attribute of an entity has the form $attr\ (=|\geq|\leq)\ value$, where $attr \in Attr_{Entity}$ and $value$ refers to an atomic value.
- **User.** A *user* u is a person who owns a resource and controls access to it. U denotes the set of all users.
- **Subject.** A *subject* s is a person who receives an authorization. A single authorization can have multiple subjects. S is the set of all subjects. $Attr_S$ stands for the set of attributes of the subjects. Given a set of attributes of a subject $Attr_S = \{attr_1^S, attr_2^S, \cdots, attr_n^S\}$, the induced set of subjects contains all subjects $s \in S$ that fulfill $Attr_S$. For instance, the set of subjects induced by $age < 20 \land income < 20k$ contains all subjects $s \in S$ whose age is smaller than 20 and whose income is less than 20k.
- **Resource.** A *resource res* is a physical or informational unit, together with attributes, e.g., type, owned by a user u for which u controls access. $Attr_{res}$ stands for the set of attributes of the resources.
- **Operation.** An *operation op* is an action that one can invoke on a resource, e.g., *read* or *write*. Op denotes the set of all operations.
- **Decision-type.** A *decision-type dt* is a right to execute an operation. D denotes the set of all decision-types.

  ABAC enables the specification of environment conditions alongside its components. For instance, a condition like "$access - day = $ Monday" specifies that the day of access must be evaluated during a request.

2.2 ABAC Authorizations

An authorization lets one decide whether a requested access should be permitted or denied, given the values of the attributes of the subject and the resource.

Definition 1 (Authorization). *Let a user $u \in U$, a set of subject attributes $Attr_S$, a set of resource attributes $Attr_{res}$, an operation $op \in Op$, and a decision-type $dt \in D$ be given. An **authorization** A is a 5-element tuple $\langle u, Attr_S, Attr_{res}, op, dt \rangle$. The authorization A indicates that user u assigns the decision-type dt to the subjects specified by $Attr_S$ to invoke the operation op on the resources specified by $Attr_{res}$.* — *We call the set of all authorizations $\mathcal{A}$. Given an authorization A, we say that $user(A)$ assigns A to $subj(A)$, and $subj(A)$ receives authorization A from $user(A)$.*

We do not specify any environment attribute as part of an authorization to ease presentation. Next, only the resource owner can create an authorization to govern the set of allowable operations over his resource. We focus on settings where each resource has one owner, e.g., physical positions. Studying settings where a resource has multiple owners, i.e., multiparty access control models, remains future work. We now introduce further notation: $subj(A)$, $res(A)$, $op(A)$ and $dt(A)$ denote, respectively, the subjects induced by $Attr_S$ of A, the resources induced by $Attr_{res}$ of A, the operation op of A, and the decision-type dt of A. Example 4 illustrates how to express an authorization.

Example 4. Consider user Anne granting *read* access to her file *File1* for all subjects with $age \geq 18$. This authorization, denoted as A_{Anne}, can be expressed in our conceptual structure as follows: $A_{Anne} = \langle$ Anne, $age \geq 18$, $name = File1$, $read$, $allow \rangle$, where age is a subject attribute in $Attr_S$, $name$ is a resource attribute in $Attr_{res}$, $read$ is an operation in Op, and $allow$ is a decision-type in D.

3 Reciprocal Authorization

3.1 Syntax and Semantics

Existing access control models are based on the decision-types $D = \{deny, allow\}$. An *allow* authorization A uses the decision-type *allow* and states that $user(A)$ authorizes $subj(A)$ to invoke $op(A)$ on $res(A)$. A *deny* authorization uses *deny* and states that $user(A)$ forbids $subj(A)$ to invoke $op(A)$ on $res(A)$.

We expand D with a new decision-type, *recip* (reciprocal), to accommodate *reciprocal authorizations*, capturing the reciprocity phenomenon. In determining whether subject s is permitted to invoke $op(A)$ on $res(A)$, the decision hinges on the authorizations received and assigned by s, as well as the sensitivity levels of the involved resources. Resource sensitivity classification is required, although for simplicity, we assume uniform sensitivity across all resources. In Sect. 5, we extend this model to address varying resource sensitivities.

Given two authorizations A and B, we use $type(res(A)) = type(res(B))$ to indicate that the resources in both authorizations are of the same type.

Definition 2 (Reciprocal authorization). *An authorization A is said to be* ***reciprocal*** *if $dt(A) = recip$. A reciprocal authorization A states that user(A) permits invoking $op(A)$ on $res(A)$ to the subjects in $subj(A)$ who have issued an authorization B to user(A), if the following boolean expression yields true: $(type(res(B)) = type(res(A))) \land ((dt(B) = allow) \lor (dt(B) = recip)) \land (op(B) = op(A))$.*

This new authorization can be integrated into existing access control models that already include *allow* and *deny* decision types, such as ABAC. We demonstrate its integration within the ABAC model, focusing on scenarios where resources are of the same type, like physical positions. Exploring scenarios with diverse resource types and sensitivities is deferred to future research.

We restrict the elements of an authorization, as follows: (1) The set of subject attributes is $Attr_S = \{age, name\}$. The set of subjects that receive a given authorization A is $subj(A) = \{s \in S \mid name=s \lor age = x, \text{where } x \in \mathbb{Z}^+\}$. (2) The resources are the positions of the users and each user $u \in U$ has one physical position, p_u. (3) The set of operations is $Op = \{read\}$. (4) The set of decision-types is $D = \{allow, recip, deny\}$.

3.2 Conflict Resolution

Definition 3 (Authorization conflict). *Given a set of authorizations $A_C \subseteq \mathcal{A}$, a subject $s \in S$ and a user $u \in U$, an **authorization conflict** exists with respect to u and s if s has received more than one authorization on the same resource with different decision-types assigned by u. An authorization conflict exists with respect to u and s if $\exists A, B \in A_C : (user(A) = user(B) = u) \land (s \in subj(A) \cap subj(B)) \land (type(res(A)) = type(res(B))) \land (dt(A) \neq dt(B))$.*

Example 5. Consider a subject s with attribute $age(s) = 18$, and the authorizations $A = \langle u, age=18, p_u, read, recip \rangle$ and $B = \langle u, name = s, p_u, read, deny \rangle$. A and B are in conflict with respect to u and s. A assigns a *reciprocal* decision-type to s while B assigns a *deny* decision-type to s for reading the same resource.

To resolve authorization conflicts, various strategies have been proposed [11], like *recency-overrides* prioritizing later specifications over earlier ones. Alternatively, we adopt a *deny-recip* precedence strategy, which prioritizes authorizations based on their decision-types.

Definition 4 (Deny-recip precedence strategy). *A deny- recip precedence strategy is a prioritization of the decision-types in D which states that a deny authorization precedes a reciprocal one and a reciprocal one precedes an allow one. We write deny $\gg$ recip $\gg$ allow.*

We prioritize precedence as *deny* $\gg$ *recip* $\gg$ *allow* to mitigate potential leakage risks [11]. Given two authorizations A and B, we write $A \gg B$ to denote that $dt(A) \gg dt(B)$. We interpret an operation not granted explicitly as denied. The conflict resolution process is as follows: Given a set of authorizations $\mathcal{B}$ and a subject s, we group all authorizations $A \in \mathcal{B}$ where $s \in subj(A)$ by the user who assigned them using an *authorization-grouping function*. Each group contains authorizations from the same user, ensuring no duplicate sets. The authorization with the highest precedence is then selected from each group. Formally:

Definition 5 (Authorization-Grouping Function). *An **authorization grouping function** $\gamma : \mathcal{P}(\mathcal{A}) \times S \to \mathcal{P}(\mathcal{P}(\mathcal{A}))$ takes as input a set of authorizations $\mathcal{B} \subseteq \mathcal{A}$ and a subject $s \in S$ and outputs a set $\mathcal{C}$ of sets of authorizations such that:*

1. $\bigcup_{\mathcal{D} \in \mathcal{C}} \mathcal{D} = \{A \in \mathcal{B} \mid s \in subj(A)\}$.
2. $\forall \mathcal{D}_1, \mathcal{D}_2 \in \mathcal{C} : \mathcal{D}_1 \neq \mathcal{D}_2 \Rightarrow \mathcal{D}_1 \cap \mathcal{D}_2 = \emptyset$.
3. $\forall \mathcal{D} \in \mathcal{C}, \forall A, B \in \mathcal{D} : user(A) = user(B) \wedge s \in subj(A) \cap subj(B)$.
4. $\forall \mathcal{D}_1, \mathcal{D}_2 \in \mathcal{C}, \forall A \in \mathcal{D}_1, \forall B \in \mathcal{D}_2 : \mathcal{D}_1 \neq \mathcal{D}_2 \Rightarrow user(A) \neq user(B)$.

Example 6. Consider the authorizations A, B and the subject s from Example 5 and the authorizations $\mathcal{B} = \{A, B, C, D, E\}$, where $C = \langle v, name=s, p_v, read, allow \rangle$, $D = \langle u, name = v, p_u, read, recip \rangle$ and $E = \langle v, age = 18, p_v, read, recip \rangle$. The output of the authorization-grouping function $\gamma(\mathcal{B}, s)$ is the set $\mathcal{C} = \{\{A, B\}, \{C, E\}\}$, where $\mathcal{C}$ contains all authorizations with subject s, and each set in $\mathcal{C}$ has authorizations assigned by the same user.

A *user-decision* tuple is a tuple $\langle t_{user}, t_{dt} \rangle$ where t_{user} is a user in U and t_{dt} is a decision-type in D. Our resolve-conflicts function does not consider the resources involved in the authorizations in conflict because we restrict our study to a specific resource, the physical positions of users.

Definition 6 (Resolve-Conflicts Function). *A **resolve-conflicts function** — $Rc : \mathcal{P}(\mathcal{A}) \times S \to \mathcal{P}(U \times D)$ is a function that takes as input a set of authorizations $\mathcal{B} \subseteq \mathcal{A}$ and a subject $s \in S$ and outputs a set $\mathcal{C}$ of user-decision tuples. For each set of authorizations $\mathcal{B}_1 \subseteq \mathcal{B}$ w.r.t. subject s that are in conflict, $\mathcal{C}$ contains a user-decision tuple $\langle t_{user}, t_{dt} \rangle$, where t_{user} is the user who has assigned the authorizations in $\mathcal{B}_1$, and t_{dt} is the decision-type with the highest precedence in $\mathcal{B}_1$ that t_{user} has given to s. Given a tuple $t \in Rc(\mathcal{B}, s)$, we use t_{user} and t_{dt} to refer to the first and second element of t, respectively. Formally,*

$$Rc(\mathcal{B}, s) = \begin{cases} \{\langle user(C), dt(C) \rangle \mid \\ \quad C \in \mathcal{B}, \forall A \in \mathcal{B} : C \gg A\} & \text{if } \forall A, B \in \mathcal{B} : user(A) = user(B) \wedge \\ & \quad s \in subj(A) \cap subj(B) \\ \bigcup_{\mathcal{B}_1 \in \gamma(\mathcal{B},s)} Rc(\mathcal{B}_1, s) & \text{otherwise} \end{cases}$$

Example 7. Consider the authorizations A, B from Example 5. To resolve conflicts, we invoke the function $Rc(\{A, B\}, s)$. Since $user(A) = user(B) \wedge s \in subj(A) \cap subj(B)$ and $deny \gg recip$, $Rc(\{A, B\}, s)$ outputs the set $\{\langle u, deny \rangle\}$.

3.3 Authorized Access Request

In essence, authorization is about asking: "Can s perform op on res_u owned by u?" This query, known as an access request, becomes crucial in reciprocal authorization contexts, where permissions assigned by both u to s and vice versa must be considered.

Definition 7 (Access request). *An **access request** $Req = \langle s, op, res_u \rangle$ is a tuple consisting of a subject s, an operation $op \in Op$ and a resource owned by a user u, res_u. An access request indicates that s requests to perform the operation op on the resource res_u owned by u.*

An access request $\langle s, op, res_u \rangle$ is authorized if, after resolving conflicts with respect to s, there exists (1) a tuple with the decision-type *allow* or (2) a tuple with the decision-type *recip*, and after resolving conflicts with respect to u there is a tuple either with the decision-type *allow* or *recip*. Formally:

Definition 8 (Authorized access request). *Given the set of authorizations $\mathcal{A}$, an **access request** $\langle s, op, res_u \rangle$ is **authorized** if one of the following conditions is met:*

1. $\exists t \in Rc(\mathcal{A}, s) : t_{user} = u \land t_{dt} = allow$
2. $\exists t \in Rc(\mathcal{A}, s), \exists e \in Rc(\mathcal{A}, u) : t_{user} = u \land t_{dt} = recip \land e_{user} = s \land (e_{dt} = allow \lor e_{dt} = recip)$.

We refer to the process of determining access authorization as the *access decision*. In ABAC, the *Policy Decision Point (PDP)* is responsible for making access decisions. While our focus has been on the concept of reciprocity, we have primarily examined scenarios where all resources share the same sensitivity. However, employing *reciprocal* authorizations introduces various contextual challenges discussed in Sect. 4, along with potential solutions.

4 Addressing Reciprocal Authorizations Issues

The concept of *reciprocal* authorizations requires that access to my resource is granted only if I can access yours, ensuring fair resource exchange. Fairness in these exchanges is measured by equitable distribution of payoffs [3]. Two principles support this fairness: profit guarantees, ensuring users benefit from exchanges, and equitable payoffs, balancing exchanged benefits. Challenges arise regarding authorization updates and resources with varying sensitivity, impacting these principles.

For simplicity, we assume each user owns a single resource, typical in contexts like health records or user positions. The techniques to address these issues are expected to extend to scenarios with multiple user resources, a topic for future exploration.

4.1 Revocation Fraud

Problem. Users may temporarily adjust their authorizations to observe others, for example, changing a *deny* to *reciprocal*, then reverting to *deny* after accessing the resources. We call this problem *revocation fraud*.

Solution. To ensure consistent benefits from reciprocal authorizations, we propose enhancing the access control model by tracking users accessing resources through them. This entails incorporating a mechanism within the PDP to monitor such access. Specifically, we suggest utilizing an access control list (ACL) for each resource, which contains subjects authorized to perform operations on that resource.

Definition 9 (Access control entry). *An access control entry is a tuple t of the form $\langle s, op \rangle$ where $s \in S$ and $op \in Op$. We use $subject(t)$ and $op(t)$ to denote, respectively, the subject and operation of t.*

Definition 10 (Access control list). *Given a resource res, the access control list of res, acl_{res}, is a list of access control entries.*

Definition 11 (Access control list: Semantics). *Given an access control list of a resource res, acl_{res}, a tuple $t \in acl_{res}$ indicates that the $subject(t)$ is permitted to perform the operation $op(t)$ on the resource res.*

Having the concept of an access control list, we propose to apply the following changes to the access control model to overcome the revocation fraud problem:

- *Registering:* If s accesses res_u using a *reciprocal* authorization, u is added to the access control list of res_s.
- *Verifying:* When u wants access to res_s, the PDP checks if u is in acl_{res_s}. If so, u is granted access to res_s regardless of the authorizations in $\mathcal{A}$.
- *Deleting:* The PDP deletes u from acl_{res_s} after u has accessed res_s.

Users cannot modify the access control lists of their resources. The PDP has to guarantee the integrity of these lists. Note that the access control list of a resource grows linearly with the number of users in the worst-case scenario where a user commits revocation fraud against all users.

According to our definition of *reciprocal* authorizations, v gains access to u's resource because both resources shared the same sensitivity level during u's access request. However, if the sensitivity of the accessed resource changes, v may access a resource with a different sensitivity level. To address this, we can set a time limit, like 24 h, for v to access u's resource, with access control entries automatically removed after this period. Implementing temporal constraints, similar to those in temporal access control models [12], ensures resource exchanges occur within a brief timeframe, preserving the original sensitivity level evaluated during the access request.

4.2 Sensitivity Problem

Problem. We have primarily examined one type of resource with uniform sensitivity, illustrated by our focus on physical positions. However, in different scenarios, resources may vary in sensitivity levels. In such cases, while *reciprocal* authorizations dictate that only resources with matching sensitivities can be exchanged, assigning sensitivity levels to resources by users can lead to unfair exchanges. We refer to this challenge as the *sensitivity problem*.

Solution. To address the sensitivity problem initially, one might consider having the system or a trusted authority assign the sensitivity level to all resources. This solution could involve extending the semantics of *reciprocal* authorizations to stipulate that users can only exchange resources with matching sensitivities. However, entrusting an entity other than the resource owner to determine the sensitivity level can be problematic, as illustrated in Example 8.

Example 8. Consider a library management system and users Anne and Bob, who aim to exchange their theses. The system assigns a degree of sensitivity to all resources. While the system may assign similar sensitivity levels to both documents because they are of the same type, Anne may perceive her thesis as more sensitive than Bob's, leading to potential fairness concerns.

Our solution to alleviate the sensitivity problem is as follows. First, owners subjectively specify the sensitivity level of their resources. Second, users set a minimum trust level for subjects accessing their resources with reciprocal authorizations. Third, a user's trust rating is computed based on feedback from users who have accessed their resources, reflecting their perception of sensitivity compared to the owner's assignment. Implementation involves defining a sensitivity scale, extending reciprocal authorization definitions (Sect. 3), and employing a reputation system. In principle, one can use any reputation system that fulfills the following requirements:

- *Score:* It has to provide a trust score for each user, a real number.
- *Rating:* It must support positive and negative ratings within a numerical scale, e.g., if u accesses a resource owned by v, res_v, with sensitivity 10, u may rate the transaction as negative if u deems the sensitivity of res_v lower.
- *Scope:* It must compute a global trust score for each user u, considering the opinions of all the peers who have interacted with u.
- *Integrity:* The system has to guarantee that users will not be able to tamper with their reputation.

The proposed approach, akin to reputation systems, may seem vulnerable to whitewashing attacks, where users with low reputations change identities to reset. However, this can be countered by existing reputation system solutions, such as limiting the number of identifiers per user, binding identifiers to IP addresses, or requiring entry fees. Next, our solution here helps mitigate sensitivity concerns and unfair behaviors like resource forgery. However, there remains a risk of information leakage if users act dishonestly. Research has shown that achieving fair resource exchange in distributed systems without a trusted third party (TTP) is challenging [13], akin to our scenario. Fully addressing sensitivity issues would require integrating a TTP into our proposed architecture and employing methods like smart contracts [14] or blockchain-based protocols [15]. However, incorporating these approaches extends beyond the scope of this article.

5 Extending Reciprocal Authorizations

In this section, we extend our definitions related to *reciprocal* authorizations by incorporating our solutions to the revocation fraud and sensitivity problems.

5.1 Trust Based Authorizations

To address sensitivity, we enhance reciprocal authorizations by enabling users to set a minimum trust level for accessing their resources. We utilize a reputation system to assess each user's trustworthiness based on evaluations from others. Denoting a user's trust value as $trust_w$, we refine the definition of authorization in Definition 1 to include user trust values.

Definition 12 (Trust-based authorization: Syntax). *Let a user $u \in U$, a set of subject attributes $Attr_S$, a set of resource attributes $Attr_{res}$, an operation $op \in Op$, a decision-type $dt \in D$, and a trust value $tr \in \mathbb{R}_{>0}$ be given. A **trust-based authorization** A_t is a 6-element tuple $\langle u, Attr_S, Attr_{res}, op, dt, tr \rangle$. We use $trust(A_t)$ to denote the trust value specified in A_t. $\mathcal{A}_T$ denotes the set of all trust-based authorizations.*

A trust-based authorization A_t indicates that user u assigns the decision-type dt to the subjects specified by $Attr_S$ to invoke the operation op on the resources specified by $Attr_{res}$ if the trust value of s, $trust_s$, is equal or greater than tr. Definition 12 includes *allow*, *reciprocal*, and *deny* authorizations. Since the sensitivity problem is only related to *reciprocal* authorizations, in the case of *allow* and *deny* authorizations, the trust value has to be ignored.

Next, we define the semantics of trust-based *reciprocal* authorizations. Given a resource res, let $sensitivity(res)$ denote the degree of sensitivity of res, i.e., the *sensitivity* value corresponding to res.

Definition 13 (Trust-based *reciprocal* authorization: Semantics). *Let a trust-based reciprocal authorization A_t, where $dt(A_t) = recip$, be given. A_t states that $user(A_t)$ permits invoking $op(A_t)$ on $res(A_t)$ to all subjects $s \in subj(A)$ who have issued an authorization B to $user(A_t)$ if the following expression evaluates to true: $(sensitivity(\ res(B_t)) = sensitivity(res(A_t))) \wedge (trust_s \geq trust(A_t)) \wedge (trust_{user(A_t)} \geq trust(B_t)) \wedge ((dt(B_t) = allow) \vee (dt(B_t) = recip)) \wedge (op(B_t) = op(A_t))$.*

Similarly to authorizations without trust conditions, authorization conflicts may arise when using trust-based authorizations. In the next section, Sect. 5.2, we discuss these conflicts and explain how to handle them.

5.2 Trust-Based Conflict Resolution

Regarding authorization conflicts in trust-based authorizations, two sources exist. First, akin to authorizations without trust, conflicts arise when a subject receives trust-based authorizations from the same user with different decision-types. Second, conflicts occur when a subject receives trust- based authorizations from the same user with identical decision-types but varying trust values.

To resolve conflicts, we prioritize authorizations according to the precedence order $deny \gg recip \gg allow$, as explained in Sect. 3. Additionally, we give precedence to trust-based authorizations with the highest trust values to mitigate the risk of potential leaks. To simplify the process, we split our conflict resolution function into two parts: one resolves conflicts based on decision types (Definition 14), while the other considers trust values (Definition 15). The decision-trust resolve-conflict function is implemented in Algorithm 1.

Given two trust- based authorizations A_t and B_t, we write $A_t \gg_{dt} B_t$ and $A_t \gg_t B_t$ to denote that $dt(A_t) \gg dt(B_t)$ and $trust(A_t) \geq trust(B_t)$.

Definition 14 (Decision Resolve-Conflicts Function). *A decision resolve-conflicts function — $Rc_{dt} : \mathcal{P}(\mathcal{A}_T) \times S \to \mathcal{P}(\mathcal{A}_T)$ is a function that takes as input a set of authorizations $\mathcal{B} \subseteq \mathcal{A}_T$ and a subject $s \in S$ and outputs a set $\mathcal{C}$ of trust- based authorizations. For each set of authorizations $\mathcal{B}_1 \subseteq \mathcal{B}$ with respect to subject s that are in conflict, $\mathcal{C}$ contains a trust- based authorization C_t such that C_t is the authorization with highest decision-type precedence in $\mathcal{B}_1$. Formally,*

$$Rc_{dt}(\mathcal{B}, s) = \begin{cases} \{C_t \mid C_t \in \mathcal{B}, & \text{if } \forall A_t, B_t \in \mathcal{B} : user(A_t) = user(B_t) \\ \forall A_t \in \mathcal{B} : C_t \gg_{dt} A_t\} & \wedge s \in subj(A_t) \cap subj(B_t) \\ \bigcup_{\mathcal{B}_1 \in \gamma(\mathcal{B}, s)} Rc_{dt}(\mathcal{B}_1, s) & \text{otherwise} \end{cases}$$

Definition 15 (Decision-Trust Resolve-Conflicts Function). *A decision- trust resolve- conflicts function — $Rc_{dt}^{tr} : \mathcal{P}(\mathcal{A}_T) \times S \to \mathcal{P}(\mathcal{A}_T)$ is a function that takes as input a set of authorizations $\mathcal{B} \subseteq \mathcal{A}_T$ and a subject $s \in S$ and outputs a set $\mathcal{C}$ of trust- based authorizations. For each set of authorizations $\mathcal{B}_1 \subseteq \mathcal{B}$ with respect to subject s that are in conflict, $\mathcal{C}$ contains a trust- based authorization C_t such that $dt(C_t)$ is the decision-type with the highest precedence in $\mathcal{B}_1$, and $trust(C_t)$ is the highest trust value in $\mathcal{B}_1$. Formally,*

$$Rc_{dt}^{tr}(\mathcal{B}, s) = \begin{cases} \{C_t \mid C_t \in \mathcal{B}, & \text{if } \forall A_t, B_t \in \mathcal{B} : \\ \forall A_t \in \mathcal{B} : C_t \gg_t A_t\} & user(A_t) = user(B_t) \wedge \\ & s \in subj(A_t) \cap subj(B_t) \\ & \wedge dt(A_t) = dt(B_t) \\ \bigcup_{\mathcal{B}_1 \in \gamma(Rc_{dt}(\mathcal{B}, s), s)} Rc_{dt}^{tr}(\mathcal{B}_1, s) & \text{otherwise} \end{cases}$$

5.3 Trust Based Authorized Access Request

Here, we define how the PDP makes the access decisions. Such a definition has to include our solutions to the revocation fraud and sensitivity problems.

Algorithm 1: decisionTrustResolveConflicts

Input : Authorization Set $\mathcal{B}_T \subseteq \mathcal{A}_T$, subject $s \in S$
Output: Authorization Set $\mathcal{C}_T \subseteq \mathcal{B}_T$
1 Initialize: $\mathcal{C} \leftarrow \emptyset$, $\mathcal{D} \leftarrow \emptyset$;
  /* Group the authorizations in $\mathcal{B}_T$ assigned to s by their assigning users and output the resulting set.   */
2 $\mathcal{D}_T \leftarrow \gamma(\mathcal{B}, s)$;
3 **foreach** *set* $\mathcal{D}_1$ *in* $\mathcal{D}_T$ **do**
4   Create $B_t \leftarrow \langle\,\rangle$;   // Create an empty authorization
5   **foreach** *authorization* A_t *in* $\mathcal{D}_1$ **do**
6   **if** $dt(A_t) \gg dt(B_t)$ **then**   // Solve decision-type conflicts
7 | $B_t \leftarrow A_t$;   // Replace B_t with A_t
8   **if** $dt(A_t) == dt(B_t)$ **then**
9 | **if** $trust(A_t) \gg trust(B_t)$ **then** // Solve trust values conflicts
10 | | $B_t \leftarrow A_t$;   // Replace B_t with A_t
11   Add B_t to $\mathcal{C}_T$;
12 **return** $\mathcal{C}_T$

Definition 16 (Authorized access request with acl). *Given the set of trust-based authorizations $\mathcal{A}_T$, an **access request** $\langle s, op, res_u \rangle$ is **authorized** if one of the following conditions is met:*

1. $\exists t_{acl} \in acl_{res_u} : subject(t_{acl}) = s$
2. $\exists A_t \in Rc_{dt}^{tr}(\mathcal{A}_T, s) : (user(A_t) = u) \wedge (dt(A_t) = allow)$
3. $\exists A_t \in Rc_{dt}^{tr}(\mathcal{A}_T, s), \exists B_t \in Rc_{dt}^{tr}(\mathcal{A}_T, u) : (user(A_t) = u) \wedge (\,dt(A_t) = recip\,) \wedge (trust(A_t) \leq trust_s) \wedge (user(B_t) = s) \wedge (\,(dt(B_t) = allow) \vee (\,(dt(B_t) = recip) \wedge (trust(B_t) \leq trust_u)\,)\,) \wedge (\,sensitivity(res(A_t)) = sensitivity(res(B_t))\,)$.

Example 9. Consider users u and v, each owning resources res_u and res_v respectively. Assume that after resolving conflicts regarding v and u, the decision-trust resolve-conflicts function produces A_t and B_t, respectively, where A_t is a trust-based authorization assigned by u to v and B_t is a trust-based authorization given by v to u. Suppose v seeks access to res_u. The steps, followed by the PDP, to assess whether v may access res_u are depicted in Fig. 1. According to Definition 1, the decision-type of an authorization can be *deny, allow* or *reciprocal*. For completeness, Fig. 1 assumes that any other decision-type results in denial of the access request.

6 Integrating LBS with Reciprocal Authorizations

In the remainder of this paper, we focus on the integration of LBS with *reciprocal* authorizations. Since users aim to exchange physical positions, we assume uniform sensitivity for these resources, thus omitting sensitivity degrees and utilizing authorizations without trust conditions. We start by defining the soundness principle that an algorithm in the context of LBSs and *reciprocal* authorizations should fulfill.

6.1 Soundness Criteria

An algorithm is sound if it is both *correct* and *complete* [16]. To define the soundness and completeness of an algorithm in the context of LBSs and *reciprocal* authorizations, we introduce two constraints: location and authorization.

Definition 17 (Location constraint). *A **location constraint**, LCons, is a predicate on physical positions.*

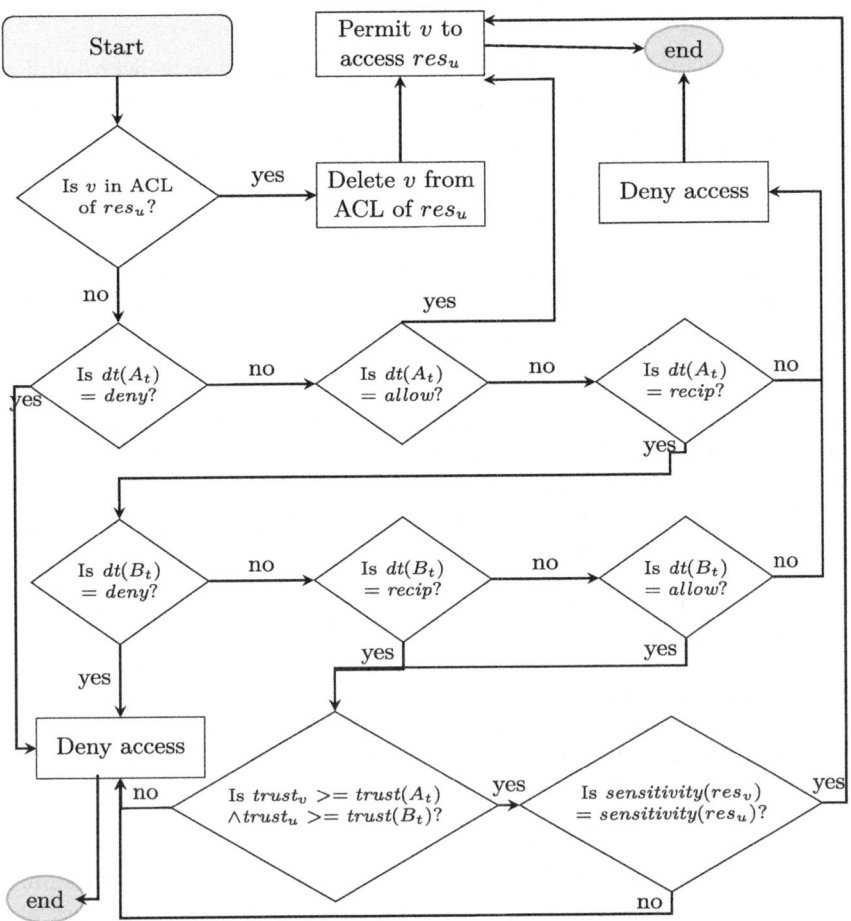

Fig. 1. PDP steps to make the access decision for Example 9

Let $dist(p_x, p_s)$ denote the distance between the physical positions of users x and s. Consider a distance d and the physical positions of two persons u and s, p_u and p_s, respectively, $dist(p_u, p_s) \leq d$ is a location constraint.

Definition 18 (Authorization constraint). *Given two persons, u and s, an **authorization constraint** ACons is a predicate on a set of authorizations M that involve persons u and s.*

Let a set of authorizations M and two persons u and s be given. The access request $\langle s, read, p_u \rangle$ is an authorization constraint. If $\langle s, read, p_u \rangle$ is authorized, the predicate evaluates to *true*; otherwise it evaluates to *false*.

Definition 19 (Query). *Given a set of persons $\mathcal{P}$, a **query** $Q(C)$ is a set of location and authorization constraints C. Its output is the elements of $\mathcal{P}$ that fulfill C. $Ans_\mathcal{P}(Q(C))$ is the output of $Q(C)$.*

A user algorithm is an algorithm that outputs a set of users who fulfill a set of constraints given as algorithm input.

Definition 20 (User algorithm). *A **user algorithm** $\Pi : \mathscr{P}(U) \times \mathscr{P}(Cons) \to \mathscr{P}(U)$ is an algorithm that has as input a set of users $U_1 \in \mathscr{P}(U)$ and a set of constraints C and outputs a set of users U_2.*

In the context of LBSs and *reciprocal* authorizations, a user algorithm Π computes a location and an authorization constraint on the physical positions of a given set of persons P. Based on these two constraints, we define the correctness and completeness of Π.

Definition 21 (Correctness in the context of LBSs and *reciprocal* authorizations). *Let a user algorithm Π, a set $U_1 \in \mathscr{P}(U)$ and a set of constraints $C = \{LCons, ACons\}$ be given. The user algorithm Π is **correct** with respect to U_1 and C if for all $u \in \Pi(U_1, \{LCons, ACons\})$, $u \in Ans_{U_1}(Q(LCons)) \land u \in Ans_{Ans_{U_1}(Q(LCons))}(Q(ACons))$.*

Correctness is given if for all users u in the output of Π, u is a person in $U1$ that fulfills $LCons$, and u is a person in $Ans_{U_1}(Q(LCons))$ that fulfills $ACons$.

Definition 22 (Completeness in the context of LBSs and *reciprocal* authorizations). *Let a user algorithm Π, a set $U_1 \in \mathscr{P}(U)$ and constraints $C = \{LCons, ACons\}$ be given. Π is **complete** with respect to U_1 and C if for all persons u with $u \in Ans_{U_1}(Q(LCons)) \land u \in Ans_{Ans_{U_1}(Q(LCons))}(Q(ACons))$, $u \in \Pi(U_1, \{LCons, ACons\})$.*

To illustrate completeness, consider a user algorithm Π that always outputs an empty set. Then Π fulfills the correctness principle. However, Π is not useful.

Definition 23 (Soundness in the context of LBSs and *reciprocal* authorizations). *Given a user algorithm Π, a set $U_1 \in \mathscr{P}(U)$ and constraints $C = \{LCons, ACons\}$, Π is **sound** with respect to U_1 and C if (1) Π is correct with respect to U_1 and C and (2) Π is complete with respect to U_1 and C.*

6.2 Primitives and Algorithms for Reciprocal Authorizations

Depending on the services offered by a system, one may need different primitives. A primitive is a fundamental unit with a specific function that can be combined with others. In the context of LBSs, determining the subjects authorized to access a user's physical position is essential for query responses. The primitives outlined below fulfill this requirement, as demonstrated in Sect. 6.4:

- *Primitive-Request:* Given two persons u and s, may s read the physical position of u?
- *Primitive-View:* Given a person s, whose physical positions is s permitted to read? We call this set $View_s$, the view of s.

Implementing these primitives requires a resolve-conflicts algorithm. In our technical report [17], we provide a simple implementation of our resolve conflict function, Definition 6, and prove its correctness. We call our algorithm $resolveConflicts(\mathcal{B}, \mathcal{U}, \mathcal{S})$, where $\mathcal{B} \subseteq \mathcal{A}$, $\mathcal{U}$ is a set of users, and $\mathcal{S}$ is a set of subjects. Our *resolveConflicts* algorithm outputs a map $autMap$, which stores pairs of keys and values. The key is a pair consisting of a user $u \in \mathcal{U}$ and a subject $s \in \mathcal{S}$, and the value of the map is the decision-type of the authorization $A \in \mathcal{B}$ with the lower precedence with respect to u and s.

In the following, we present our algorithms, *Pr-Request* algorithm and *Pr-View* algorithm to implement the primitives *Primitive-Request* and *Primitive-View*, respectively. Since authorization conflicts can exist, both algorithms make use of the *resolveConflicts* algorithm.

Given two persons u and s, Algorithm 2 determines if s can read the physical position of u. Algorithm 2 starts by initializing, among others, the set Set_{req}, containing the access requester, and the set Set_{pp}, containing the person of the requested physical position. executes the *resolveConflicts* algorithm on sets $\mathcal{A}$, Set_{pp}, and Set_{req}, storing the result in map $ReceiveAut_s$. As Set_{pp} and Set_{req} have only one element each, $ReceiveAut_s$ comprises a single entry e with key (u, s), representing the decision-type of the highest-precedence authorization assigned by u to s. If $e.getValue() = allow$, the algorithm returns $true$. If $e.getValue() = recip$, the algorithm invokes the *resolveConflicts* algorithm on the sets $\mathcal{A}$, Set_{req} and Set_{pp}, and stores the output in the map $ReceiveAut_u$, Line 7. $ReceiveAut_u$ contains only one entry t with key (s, u), representing the decision-type of the highest-precedence authorization assigned by s to u. If $t.getValue() = allow$ or $t.getValue() = recip$, the algorithm returns $true$, indicating s is permitted to access p_u. Otherwise, access is denied for s.

Given a person s, Algorithm 3 outputs the view of s. Algorithm 3 computes the view of a given person s. It initializes sets $\mathcal{U}$ and $\mathcal{S}$, assigning the set of all users U to $\mathcal{U}$ and s to $\mathcal{S}$. The algorithm also invokes the *resolveConflicts* algorithm on the sets $\mathcal{A}$, $\mathcal{U}$ and $\mathcal{S}$, and stores the output in the map $ReceiveAut_s$. For each entry e in $ReceiveAut_s$, if $e.getValue() = allow$, the user represented by $e.getKey.User()$ is added to set $View_s$ (Lines 3–4). If $e.getValue() = recip$, $e.getKey.User()$ is added to the set $RecipRA$ (Lines 5–6). If $RecipRA$ is not empty, then the algorithm invokes the *resolveConflicts* algorithm on the sets $\mathcal{A}$,

S and *RecipRA*, and stores the output in the map $AuthMap_s$. For each user $u \in RecipRA$, it checks if $AuthMap_s$ contains an entry with key (s, u) and a value of *allow* or *recip*. If such an entry exists, u is added to set $View_s$.

Algorithm 2: Pr-Request

   Input : Authorization Set $\mathcal{A}$, Access request $\langle s, read, p_u \rangle$
   Output: Boolean resp
1. Initialize: $Set_{pp} \leftarrow \{u\}$, $Set_{req} \leftarrow \{s\}$,
   $ReceiveAut_s \langle (user, subject), dt \rangle \leftarrow$ *empty map*,
   $ReceiveAut_u \langle (user, subject), dt \rangle \leftarrow$ *empty map*;
2. $ReceiveAut_s \leftarrow resolveConflicts(\mathcal{A}, Set_{pp}, Set_{req})$;
3. **foreach** *entry e in ReceiveAut$_s$* **do**
4.   **if** $e.getValue() = allow$ **then return** true
5.   **if** $e.getValue() = recip$ **then**
6. $ReceiveAut_u \leftarrow resolveConflicts(\mathcal{A}, Set_{req}, Set_{pp})$;
7.   **foreach** *entry t in ReceiveAut$_u$* **do**
8.   **if** $t.getValue() = allow \lor t.getValue() = recip$ **then**
9.   **return** true;
10. **return** false ;

Algorithm 3: Pr-View

   Input : Authorization Set $\mathcal{A}$, person s
   Output: Set $View_s$
1. Initialize: $\mathcal{U} \leftarrow U$, $\mathcal{S} \leftarrow \{s\}$, $View_s \leftarrow \emptyset$,
   $ReceiveAut_s \langle (user, subject), dt \rangle \leftarrow resolveConflicts(\mathcal{A}, \mathcal{U}, \mathcal{S})$,
   $AA_s \langle (user, subject), dt \rangle \leftarrow$ *empty map*, $RecipRA \leftarrow \emptyset$;
2. **foreach** *entry e in ReceiveAut$_s$* **do**
3.   **if** $e.getValue() = allow$ **then**
4.   add $e.getKey.User()$ to $View_s$;
5.   **if** $e.getValue() = recip$ **then**
6.   add $e.getKey.User()$ to $RecipRA$;
7. **if** $RecipRA.size() \neq 0$ **then**
8. $AuthMap_s \leftarrow resolveConflicts(\mathcal{A}, \mathcal{S}, RecipRA)$;
9.   **foreach** *u in RecipRA* **do**
10.   **if** $AuthMap_s.get((s, u)) = recip \lor AuthMap_s.get((s, u)) = allow$ **then**
         add u to $View_s$
11. **return** $View_s$;

Next, we prove that given the authorization set $\mathcal{A}$ and a person s, the output of the *Pr-View* algorithm, Algorithm 3, contains all persons that s is permitted to read their physical positions and not more. The proof of Algorithm 2 can be done in similar way as the proof of Algorithm 3.

Lemma 1. *Given an authorization set $\mathcal{A}$ and a person s,*

1. *For all persons $u \in Pr\text{-}View(\mathcal{A}, s)$, Algorithm 3, s is authorized to read the physical position of u with respect to Definition 8.*
2. *If $u \notin Pr\text{-}View(\mathcal{A}, s)$, then s is not authorized to read the physical position of u with respect to Definition 8.*

Proof. We will prove that for each user $u \in Pr\text{-}View(\mathcal{A}, s)$, $\langle s, read, p_u \rangle$ is authorized, Definition 8, if either condition (1) or (2) is met:

(1) $\exists t \in Rc(\mathcal{A}, s) : t_{user} = u \land t_{dt} = allow$.
(2) $\exists t \in Rc(\mathcal{A}, s), \exists l \in Rc(\mathcal{A}, u) : t_{user} = u \land t_{dt} = recip \land l_{user} = s \land (l_{dt} = allow \lor l_{dt} = recip)$

In our technical report [17], we prove that our *resolveConflicts* algorithm is correct, Definition 6. That is, for each authorization in conflict with respect to a subject s, its output contains an entry of the form $e = ((t_{user}, s), t_{dt})$, where t_{user} refers to a user $u \in U$ and t_{dt} is the decision-type with the highest precedence with respect to t_{user} and s. Example 7 illustrates the output of our resolve conflict function. Then in conditions (1) and (2), it is possible to replace a tuple $t = \langle t_{user}, t_{dt} \rangle \in Rc(\mathcal{A}, s)$ with an entry $e = ((t_{user}, s), t_{dt}) \in resolveConflicts(\mathcal{A}, \{U\}, \{s\})$. First, a person u is added to $Pr\text{-}View(\mathcal{A}, s)$, if $t_{user} = u \land t_{dt} = allow$, Lines 3–4. That is, this step evaluates condition (1). Second, if $t_{dt} = recip$, then u is added to the set *RecipRA*, Lines 5–6. Then for each $u \in RecipRA$, u is added to $Pr\text{-}View(\mathcal{A}, s)$ if there exists an entry f in $resolveConflicts(\mathcal{A}, \{s\}, RecipRA)$ such that $f = ((s, u), allow)$ or $f = ((s, u), recip)$, Lines 10–11. That is, this step evaluates condition (2).

6.3 System Architecture

We consider a system architecture that contains a LBS provider *LBSP* comprising a PDP, a user database DB_U, storing each user u's position p_u, and an authorization database DB_A, housing the authorizations $\mathcal{A}$ (see Fig. 2). We assume a database management system featuring R-tree indexing for spatial query processing on DB_U and B-tree indexing for authorization queries on DB_A. Our prototype, developed in Java, includes the LBS and access control in charge of the PDP, employing the primitives detailed in Sect. 6.2. We concentrate on a conceptual level with a centralized server; exploring architectural design alternatives in distributed environment with multiple PDPs remains a subject of future investigation.

The *LBSP* facilitates location-dependent queries, particularly focusing on k-nearest neighbor and range queries.

Definition 24 (Location-dependent query). *Given a set of persons $\mathcal{P}$, a **location-dependent query** $Q(LCons)$ takes a location constraint LCons and outputs the persons that fulfill it.*

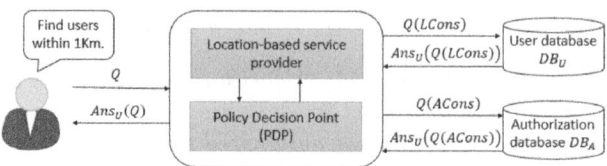

Fig. 2. System Architecture

Definition 25 (k-Nearest Neighbor query). *Given an integer k and a person s, a **k-nearest neighbor (kNN) query** $knn(k,s)$ is a location-dependent query where the location constraint $knn_{k,s}(p_u)$ is:* $\forall M \subseteq U, \big(\forall x \in M, dis(p_x, p_s) < dis(p_u, p_s)\big) \Rightarrow |M| \leq k$, *where U is the set of all users. In words, if the previous predicate evaluates to true for a physical position p_u, then the corresponding person u is in the result of the $knn(k,s)$ query; otherwise not.*

Definition 26 (Range query). *Given a distance d and a person s, a **range query** $range(d,s)$ is a location-dependent query with the following constraint $range_{d,p_s}(p_x)$: $dist(p_x, p_s) \leq d$.*

Definition 25 is a higher order logic. However, to facilitate proofs that our proposed approaches are sound, we will use a recursive definition, Definition 27.

Definition 27 (k-Nearest Neighbor query recursive definition). *A **k-nearest neighbor query** knn is a location-dependent query, where the location constraint consists of two elements (k,s), where k is an integer number and s is the person who issues the query. The result $Ans(knn)$ of such a query is the set of users $u \in U$ such that $|Ans(knn)| = k \wedge \forall u \in Ans(knn), \forall v \in U \setminus Ans(knn): dist(p_s, p_u) \leq dist(p_s, p_v)$.*

Definition 28 (Bounded result-size query). *Given a location- dependent query $Q(LCons)$, Q is a **bounded result-size query** if the location constraint ($LCons$) contains an explicit restriction on the number of elements of $Ans(Q)$. Otherwise, Q is an unbounded result-size query.*

kNN and range queries represent bounded and unbounded result-size queries, respectively. For a bounded result-size query Q, filtering the query result $Ans(Q)$ for users whose position s is authorized to access may lead to non-compliance with the original constraint $LCons$. We refer to the *authorizations received* by s as the set containing all authorizations $A \in \mathcal{A}$ where $s \in subj(A)$.

Example 10. Consider a kNN query $knn = (2, s)$. Suppose that (1) the neighbors of s are u, v and w, and their distances to s are 1, 2 and 3 km, respectively, and (2) s has received two authorizations A, B where $user(A) = v$, $user(B) = w$ and $dt(A) = dt(B) = allow$. The LBSP evaluates the kNN query and outputs $Ans(kNN) = \{u, v\}$. After filtering $Ans(kNN)$ based on the authorizations s has received, the result contains only $\{v\}$, which does not meet the constraint $k = 2$. The parameter should have been set to $k = 3$, to obtain $\{v, w\}$ after the filtering.

Example 11. Continuing with Example 10, consider a range query $range(2km, s)$. The LBSP evaluates the query and outputs $Ans(range) = \{u, v\}$. After filtering $Ans(range)$ w.r.t. the authorizations that s has received, the filtered result is $\{v\}$, similar to Example 10. This result fulfills the constraint $range(2km, s)$.

6.4 Design Alternatives for Integrating LBSs with Reciprocal Authorizations

To integrate *reciprocal* authorizations in the system architecture, we see two design alternatives, called *Querying-Filtering (QF)* and *Filtering-Querying (FQ)*. QF has the advantage that it can leverage existing LBS implementations. However, it has some limitations that could affect the performance, like the need to restart querying, as we will discuss. FQ does not have this need to restart.

Querying-Filtering Approach (QF). Given a location-dependent query $Q(LCons)$, the QF approach works as follows: (1) The LBSP executes Q on the user database DB_U and returns $Ans(Q)$. (2) It filters $Ans(Q)$ for the persons whose position s may read. For the filtering, there are two options:

(a) Verify for each person $u \in Ans(Q)$ if $\langle s, read, p_u \rangle$ is authorized, i.e., execute *Primitive-Request*. If so, then u is added to the final answer.
(b) Compute *Primitive-View*, $View_s$. The final answer is the intersection of $View_s$ and $Ans(Q)$.

In option (a), addressing each person in $Ans(Q)$ necessitates reading all authorizations in $\mathcal{A}$ to resolve authorization conflicts. However, with option (b), although conflicts still need resolution, elements of $\mathcal{A}$ are read at most twice. Initially, the algorithm retrieves authorizations assigned to the querying person. In the next, we focus on QF only in combination with (b).

Algorithms 4 and 5 show the details for implementing QF for kNN and range queries, respectively. Our algorithms assume that the LBSP uses existing spatial search algorithms to answer location-dependent queries, like B-tree or R-tree. We call the services used by LBSP to compute kNN and range queries, *computeKNN(knn)* and *computeRange(range)*, respectively, where knn and $range$ are the query location constraints. Next, we describe both algorithms.

Algorithm 4 requires the set of authorizations $\mathcal{A}$, the kNN query parameters k, and a person s as input. As kNN queries are bounded result-size queries, ensuring the final result complies with the query parameter k is essential. The algorithm utilizes a *while* loop to encapsulate the querying and filtering process. Within this loop, it invokes a function $estimateK(k, k_{old}, s)$ (Line 4) to estimate an integer value $k_{all} \geq k$ such that after computing the k_{all}-nearest neighbors of s and filtering based on $View_s$, the filtered result satisfies the constraint (k, s). The algorithm computes the k_{all}-nearest neighbors of s in ascending distance order and stores the result in $temp_{all}$ (Line 5). Then, for each $u \in temp_{all}$, if $u \in View_s$, u is added to $temp_{visible}$, preserving the distance order. If $temp_{visible}$ has at least k elements, the algorithm outputs the first k. Otherwise, it updates the parameter k_{old} with the current value of k_{all}, restarting the querying and filtering process. In each iteration, *estimateK* calculates a new integer value $k_{all}>k_{old}$.

Algorithm 5 receives as input the set of authorizations $\mathcal{A}$, a distance parameter *dist* and a person s. Algorithm 4 invokes Algorithm 3 to compute the view of s w.r.t. the set of authorizations $\mathcal{A}$. Then the algorithm runs the function *computeRange* with the parameters *dist* and s, and stores the result in the set $temp_{all}$. The final output is the intersection of the sets $temp_{all}$ and $View_s$.

Algorithm 4: Querying-Filtering kNN

Input : Authorization Set $\mathcal{A}$, int k, person s
Output: Set *Ans*
1 Initialize: $View_s \leftarrow \emptyset$, $Ans \leftarrow \emptyset$, $temp_{all} \leftarrow [\ ]$, $temp_{visible} \leftarrow [\ ]$, $notEnough \leftarrow true$, $k_{all} \leftarrow 0$, $k_{old} \leftarrow 0$;
2 $View_s \leftarrow Pr\text{-}View(\mathcal{A}, s)$;
3 **while** *notEnough* **do**
4     $k_{all} \leftarrow estimateK(k, k_{old}, s)$;
5     $temp_{all} \leftarrow computeKNN(k_{all}, s)$;
6     $temp_{visible} \leftarrow filter(temp_{all}, View_s)$;
7   **if** $temp_{visible}.size() \geq k$ **then**
8        $Ans \leftarrow$ select $topK(k, temp_{visible})$;
9        $notEnough \leftarrow false$;
10   **else** $k_{old} \leftarrow k_{all}$
11 **return** *Ans*;

Algorithm 5: Querying-Filtering Range

Input : Authorization Set $\mathcal{A}$, double *dist*, person s
Output: Set *Ans*
1 Initialize: $View_s \leftarrow \emptyset$, $Ans \leftarrow \emptyset$, $temp_{all} \leftarrow \emptyset$;
2 $View_s \leftarrow Pr\text{-}View(\mathcal{A}, s)$;
3 $temp_{all} \leftarrow computeRange(dist, s)$;
4 $Ans \leftarrow temp_{all} \cap View_s$;
5 **return** *Ans*;

Filtering-Querying Approach (FQ). The FQ approach works as follows, given a location-dependent query $Q(LCons)$: (1) The LBSP invokes the *Pr-View* algorithm to determine the persons whose positions s is permitted to see (*Primitive-View*). (2) The LBSP executes Q over these persons and outputs a final result. In contrast to QF, where pre-computed materializations like *computeKNN(knn)* cannot be used since the evaluation of $Q(LCons)$ must occur on the filtered outcome, the LBSP requires new primitives to execute the supported queries. We have identified two primitives: (1) $computeD(p_s, p_u)$, which computes the distance between the physical positions of two given individuals s and u, and (2) *sort by distance* $sortByD(M)$, where M represents a list of tuples containing a person u and a distance d. For the first primitive, we employ the well-known Haversine distance [18], and for the second one, we utilize the merge sort algorithm.

Algorithms 6 and 7 implement kNN and range queries, respectively. Algorithm 6 computes the view of s by invoking Algorithm 3 on $\mathcal{A}$ and s, and storing the output in $View_s$. For each person $u \in View_s$, the algorithm computes the distance between u and s, adding $\langle u, d \rangle$ to $Dist$ (Lines 4–5). Algorithm 6 sorts by distance in ascending order the set $Dist$ and for each tuple $\langle u, d \rangle$ it stores u in the list $View_{order}$. Finally, the algorithm selects the first k elements of the list $View_{order}$.

Algorithm 7 receives as input the set of authorizations $\mathcal{A}$, a distance parameter $dist$ and a person s. Steps 2–4 are the same as the ones of Algorithm 6. Then, Algorithm 7 examines each tuple $t \in Dist$ to determine if its distance $t_{distance}$ is less than or equal to $dist$. If affirmative, the individual t_{pers} stored in t is appended to the final answer (Lines 5–7).

Algorithm 6: Filtering-Querying kNN

Input : Authorization Set $\mathcal{A}$, int k, person s
Output: Set Ans
1 Initialize: $View_s \leftarrow \emptyset$, $Dist \leftarrow \emptyset$, $View_{order} \leftarrow [\ ]$, $Ans \leftarrow \emptyset$;
  $View_s \leftarrow Pr\text{-}View(\mathcal{A}, s)$;
2 **foreach** *person u in $View_s$* **do**
3 $\quad$ double $d \leftarrow computeD(p_s, p_u)$;
4 $\quad$ add tuple $\langle u, d \rangle$ to $Dist$
5 $View_{order} \leftarrow sortByD(Dist)$;
6 $Ans \leftarrow$ select $topK(k, View_{order})$;
7 return Ans

Advantages and Disadvantages of QF and FQ: With QF, LBSP can utilize existing index structures in the user database. However, in bounded result-size queries, the LBSP may need to restart the query if the filtered result fails to meet the initial constraints, cf. Example 10. With FQ, bounded result-size queries do not require restarts, Example 11. However, the evaluation of location-dependent queries must take place on the filtered result. The LBSP cannot use the index structures of the user database to execute queries efficiently. Finally, with both approaches, the costs of updates, i.e., user positions and authorizations updates, only depend on the scalability and costs of updating the index structures used. The analysis of the impact of updates is beyond the scope of this paper. It also can be found elsewhere [19].

Algorithm 7: Filtering-Querying Range

Input : Authorization Set $\mathcal{A}$, double $dist$, person s
Output: Set Ans

1 Initialize: $View_s \leftarrow \emptyset$, $Dist \leftarrow \emptyset$, $neighbors$, $Ans \leftarrow \emptyset$; $View_s \leftarrow Pr\text{-}View(\mathcal{A}, s)$;
2 **foreach** *person u in $View_s$* **do**
3 $\quad$ double $d \leftarrow computeD(p_s, p_u)$;
4 $\quad$ add tuple $\langle u, d \rangle$ to $Dist$
5 **foreach** *tuple t in Dis* **do**
6 $\quad$ **if** $t_{distance} \leq dist$ **then** $Ans \leftarrow t_{pers}$
7 return Ans

QF and FQ Are Sound. In this paper, we assume that the algorithms used to evaluate a given location-dependent query are correct and complete with respect to the location constraint $LCons$. This means that the integration of these algorithms into the context of *reciprocal* authorizations is correct and complete. The proofs that our integration of the algorithms to answer range queries into the context of *reciprocal* authorizations, Algorithms 5 and 7, are sound can be done in the same manner following the proofs of Lemmas 2 and 3. Therefore, we only present the proofs for the algorithms that support kNN queries.

Lemma 2. *Let a set of authorizations $\mathcal{A}$ and a location constraint (k, s) of a kNN query be given, where k is an integer, and s is the query issuer. Algorithm 4, QF for kNN queries, is sound.*

Proof. According to Definition 23, an algorithm is sound if it is correct and complete. Let Ans be the result output by Algorithm 4. We first prove that Algorithm 4 is correct w.r.t. Definition 21. We assume that the service used by Algorithm 4 to compute a given kNN query, $computeKNN(k_{all}, s)$ where $k_{all} \geq k$ is correct w.r.t. (k_{all}, s). If a person u is in Ans, u is in $temp_{visible}$, Line 8. If u is in $temp_{visible}$, then u is $temp_{all}$ and u is in $View_s$, Line 6. $View_s$ is the output of $Pr\text{-}View(\mathcal{A}, s)$, so s is authorized to read the physical position of u, Lemma 1. Then, $u \in rst(ACons_{\mathcal{A},s}, U)$, where $ACons_{\mathcal{A},s}$ is an authorization constraint. Since u is in $temp_{all}$, u is in $computeKNN(k_{all}, s)$. Since $computeKNN(k_{all}, s)$ is correct, then u satisfies the location constraint (k_{all}, s) w.r.t U. Furthermore, the size of $temp_{visible}$ is greater or equal than k, and $topK$ selects the k first elements from $temp_{visible}$. Then $u \in rst((k, s), rst(ACons_{\mathcal{A},s}, U))$. Consequently, Algorithm 4 is correct. Now we prove that Algorithm 4 is complete w.r.t. Definition 22. Consider a person u with $u \in rst(ACons_{\mathcal{A},s}, U) \wedge u \in rst((k, s), rst(ACons_{\mathcal{A},s}, U))$. Because u satisfies the authorization constraint w.r.t. U, u is in $View_s$. Since $View_s = rst(ACons_{\mathcal{A},s}, U)$, Definition 18, u is in $rest((k, s), View_s)$ and $rest((k, s), View_s) \subseteq rest((k_{all}, s), U)$, then u is in $rest((k_{all}, s), U)$. Next, we know that $computeKNN(k_{all}, s)$ is complete. Then u is in $temp_{all}$ and u is in $temp_{visible}$. Because $u \in rst((k, s), View_s)$, then u is in $topK$ and u is Ans. Hence, Algorithm 4 is complete; consequently, it is sound.

Lemma 3. *Let a set of authorizations $\mathcal{A}$ and a location constraint (k,s) of a kNN query be given, where k is an integer, and s is the query issuer. Algorithm 6, FQ for kNN queries, is sound.*

Proof. According to Definition 23, an algorithm is sound if it is correct and complete. Let Ans be the result output by Algorithm 6. We first prove that Algorithm 6 is correct w.r.t. Definition 21. If a person u is in Ans, u is in the list $View_{order}$, Line 6. Then there is a 2-element tuple $\langle u, d \rangle$ in $Dist$, which means u is in $View_s$. $View_s$ is the output of Pr-$View(\mathcal{A}, s)$. So s is authorized to read the position of u, Lemma 1. Then $u \in rst(ACons_{\mathcal{A},s}, U)$, where $ACons_{\mathcal{A},s}$ is an authorization constraint. Algorithm 6 uses the primitives $computeD$, $sortByD$, and $topK$ to compute the k-nearest neighbors of a given person s. We assume that the combination of these primitives to compute a given kNN query is correct w.r.t. the location constraint (k,s). Since these primitives compute the result using as input the set $View_s$, Line 2, $u \in rst((k,s), View_s)$, and $View_s = rst(ACons_{\mathcal{A},s}, U)$. Then Algorithm 6 is correct. We now prove that Algorithm 6 is complete w.r.t. Definition 22. Consider a person u with $u \in rst(ACons_{\mathcal{A},s}, U) \wedge u \in rst((k,s), rst(ACons_{\mathcal{A},s}, U))$. Because u satisfies the authorization constraint w.r.t. U, $u \in View_s$. Since $u \in View_s$, there is a 2-element tuple $\langle u, d \rangle$ in $Dist$. Then u is in $View_{order}$, Line 6. Because u satisfies the location constraint (k,s) w.r.t. the set $View_s$, u is in $topK(k, View_{order})$. Then u is in Ans. Hence, Algorithm 6 is complete. Therefore, Algorithm 6 is sound.

7 Time Complexity Analysis

A complexity analysis is helpful to predict the behavior of FQ and QF and to facilitate meaningful comparisons. An average complexity analysis depends on the internal behavior of the database, which is specific to the product and is not openly available. Furthermore, if there are changes in the system settings, the average analysis is void. So, our complexity analysis targets the worst case, which offers stronger guarantees.

7.1 Time Complexity Analysis of QF and FQ

To fulfill a given location constraint (k,s) of a kNN query, Algorithm 4 uses an estimation function $estimateK$, which estimates a value $k_{all} \geq k$, such that after computing the k_{all}-nearest neighbors of s and filtering the result based on $View_s$, the filtered result satisfies (k,s). Let k_{real} be equal to the value k_{all} of the last run of Algorithm 4. Let further be $\delta = k_{real} - k$. Next, let us assume that $estimateK$ computes the value k_{real} in the first run. We discuss this assumption later in Sect. 7.2.

Lemma 4. *Let the number of persons n, a kNN query $knn=(k,s)$, the view size of the query issuer, s, $|View_s|$, and a set of authorizations $\mathcal{A}$, be given. The time complexity of QF with no restarts is $T_C = \mathcal{O}(n + (k + \delta) \cdot |View_s|) + \mathcal{O}(\mathcal{A})$.*

Proof. The following steps are required to compute a given kNN query with the *querying-filtering* approach, with no restarts:

$Step_1$ computes the view $View_s$ of the query issuer s. We use $\mathcal{O}(\mathcal{A})$ to denote the complexity of this step.

$Step_2$ searches the $(k+\delta)$-nearest neighbors in the user database. The complexity of a kNN query using R-tree indexes is $\mathcal{O}(n)$ [20]. We validated through initial experiments that this complexity applies to the praxis.

$Step_3$ filters the result by checking for each person returned in $Step_2$ if the person is in the view $View_s$. The complexity of this step is $\mathcal{O}((k+\delta) \cdot |View_s|)$.

Consequently, the time complexity of executing a kNN query with the *QF* approach is $T_C = \mathcal{O}(n + (k+\delta) \cdot |View_s|) + \mathcal{O}(\mathcal{A})$.

Lemma 5. *Let the number of persons n, a kNN query $knn = (k,s)$, the size of the view of the query issuer s, $|View_s|$, and the set of authorizations $\mathcal{A}$ be given. The time complexity of FQ is $T_C = \mathcal{O}(|View_s| \cdot \log(n) + |View_s| + |View_s| \cdot \log(|View_s|) + k) + \mathcal{O}(\mathcal{A})$.*

Proof. The following steps are required to compute a given kNN query with the *filtering-querying* approach:

$Step_1$ computes the view, $View_s$ of the query issuer s. We use $\mathcal{O}(\mathcal{A})$ to denote the complexity of this step.

$Step_2$ looks up in the user database to obtain the physical position of each person in the view $View_s$. This has a complexity of $\mathcal{O}(|View_s| \cdot \log(n))$.

$Step_3$ computes the distance between the querying user and each of the persons in the view $View_s$. The complexity of this step is $\mathcal{O}(|View_s|)$.

$Step_4$ orders the persons in the view $View_s$ by distance to the querying user s in ascending order. The order is done using the merge sort algorithm. The complexity of this step is $\mathcal{O}(|View_s| \cdot \log(|View_s|))$.

$Step_5$ selects the k first persons. This has a complexity of $\mathcal{O}(k)$.

Consequently, the time complexity of executing a kNN query with the *FQ* approach is $T_C = \mathcal{O}(|View_s| \cdot \log(n) + |View_s| + |View_s| \cdot \log(|View_s|) + k) + \mathcal{O}(\mathcal{A})$.

We note that, since $\forall x{>}0, n{>}0 : x{>}x \cdot log(n)$, the time complexity of FQ can be further simplified to $T_C = \mathcal{O}(|View_s| + k) + \mathcal{O}(\mathcal{A})$. However, to allow a more accurate comparison of both approaches, Sect. 7.2, we do not simplify it.

7.2 Comparison of the QF and FQ Approaches

To decide which approach is better to answer a given query, one needs to compare the complexity of QF and FQ approaches and find their intersection points:

$$\mathcal{O}(n + (k+\delta) \cdot |View_s|) + \mathcal{O}(\mathcal{A}) = \mathcal{O}(|View_s| \cdot \log(n) + |View_s| + \\ |View_s| \cdot \log(|View_s|) + k) + \mathcal{O}(\mathcal{A}) \quad (1)$$

In order to solve Eq. (1) for $|View_s|$, we first omit the big-O notation:

$$n + (k + \delta) \cdot |View_s| = |View_s| \cdot \log(n) + |View_s| + |View_s| \cdot \log(|View_s|) + k \quad (2)$$

Solving Eq. (2) for $|View_s|$ yields (3). For given values of n, k, δ, (3) is the size of the view so that the time complexity in the worst case is equal. We refer to this size of the view as $View_{eq}$. $\mathcal{W}$ in (3) is the Lambert-W function [16].

$$View_{eq}(n, k, \delta) = \frac{n \cdot \ln(2) - k \cdot \ln(2)}{\mathcal{W}(2^{1-k-\delta} \cdot n \cdot (n-k) \cdot \ln(2))} \quad (3)$$

We now analyze Eq. (3) with the best-case scenario for QF, which is the one where the nearest neighbors of s are the persons whose positions s is permitted to read, i.e., $\delta = 0$. Equation (3) depends on the parameters: n, k, δ. To further simplify it, similarly to other approaches [21], we set the parameters k of the kNN query to 20, and $\delta = 0$. Then $View_{eq}$ only depends on the number of persons n.

$$View_{eq}(n, 20, 0) = \frac{n \cdot \ln(2) - k \cdot \ln(2)}{\mathcal{W}(2^{-19} \cdot n \cdot (n-20) \cdot \ln(2))} \quad (4)$$

Figure 3 plots the QF and FQ approaches, for $k = 20$ and $\delta = 0$. The x-axis is the number of persons n, the y-axis is the size of the view $|View_s|$ of the query issuer s, and the z-axis is the time complexity T_C. Figure 3 shows the intersection points of both approaches. Given an intersection point and its corresponding number of persons n, Eq. (4) yields the size of its view. We conclude that, for a given n, if $|View_s| < View_{eq}$, the time complexity of FQ is smaller than that of QF, and vice versa. In Table 1, using Eq. (4), we list the intersection points of QF and FQ, for different numbers of persons n. For instance, if $n = 2000$, $View_{eq} \approx 1014.31$. Then, for $n = 2000$, if the size of the view of the query issuer is smaller than approximately 1014.31, FQ performs better than QF.

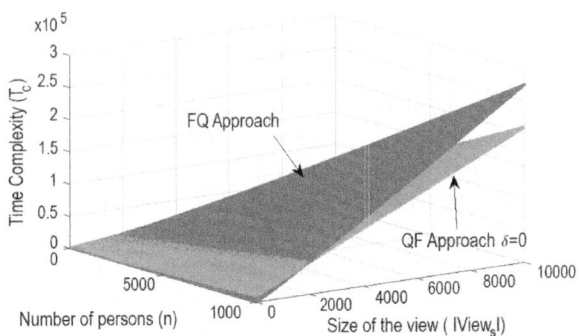

Fig. 3. Complexity of the QF and FQ Approaches– knn Query ($k = 20, \delta = 0$)

Table 1. Intersection points of the TC of QF and FQ ($k = 20, \delta = 0$)

n	2000	4000	10000	20000
$View_{eq} \approx$	1014.3096	1231.4594	1920.9585	2935.2272
n	40000	100000	317080	3000000
$View_{eq} \approx$	4709.5727	9267.9800	23032.341	152046.4307

The plot in Fig. 3 and the values in Table 1 correspond to the best case scenario for QF, i.e., $\delta = 0$. We now explain why a focus on this case is sufficient.

Let us consider real scenarios such as online social networks like Orkut and LiveJournal. The number of connections that a person s has in these networks is the number of persons that have declared to have a relationship with s. In our authorization model, this number represents the size of the view of s. Considering about 3 million nodes, the average number of connections of a person in Orkut and LiveJournal is 223.99 and 520.04, respectively [22]. In DBLP, with 317080 nodes, the average number of connections is 64.98 [22]. This suggests that the size of the view of a given person increases monotonically with the number of persons. Analogously, Table 1 reveals that $View_{eq}$ grows monotonically with the number of persons n. We can also observe that if $n = 2000$, $View_{eq}$ is greater than the average number of connections for 3 million persons in real scenarios. For n equal to 3 million, $View_{eq} = 152046$, which indicates that the size of the view of a given person in real scenarios is smaller than $View_{eq}$ for a given n. Then, FQ performs better than QF, even in the best-case scenario of QF. So, we do not dwell on the behavior of QF with restarts.

The analysis of the approaches for range queries can be done in the same way. QF for range queries does not need any restart. In the analysis of QF for kNN queries, we did not consider restarts because we assumed *estimateK* computes k_{real} in the first run. Therefore, the skeleton structure of the complexity analysis is the same for both types of queries. For these reasons, we omit this part.

8 The Size of the View of a Person

In this section, we study the impact of *reciprocal* authorizations on the view of s, $View_s$. This is important not only from the point of view of the LBSP but also from the user's perspective. Users may want to know how the use of *reciprocal* authorizations compared to the use of *deny* or *allow* in the population affects the number of persons whose position they are permitted to read.

To study the impact of *reciprocal* authorizations on the view of a person, we derive the probability $P(|View_s| = N)$ that a person s chosen at random can see the physical position of a specific number of persons N, given the share of *reciprocal* to *deny* and *allow* authorizations in the entire population. To do so, we look at the authorizations after solving all authorization conflicts. This allows us to represent the authorizations and persons as a so-called authorization graph

G. Since every person of a pair assigns an authorization to the other one, at least implicitly, G is a complete digraph.

Definition 29 (Authorization Graph). *Given a set of authorizations $\mathcal{A}$, an **authorization graph** $G = (V, E)$ is a complete digraph with labeled directed edges, as follows: The vertices represent the persons. A directed edge with label decision-type between u and v indicates that there exists a user-decision tuple $\langle u, dt \rangle \in Rc(\mathcal{A}, v)$, where $dt \in D$.*

The number of incoming edges of any node is $|E_{in}| = |V| - 1$. To compute $P(|View_s| = N)$, where $N \leq |E_{in}|$, we need a concrete distribution of *allow*, *deny* and *reciprocal* authorizations. The arguments behind our analysis hold for any distribution. For the sake of simplicity, we now assume a uniform distribution of *allow*, *deny* and *reciprocal* authorizations, where the numbers of these authorizations are parameters of the distribution, i.e., the probability that a random chosen authorization has an *allow* decision-type is given by the number of *allow* authorizations over the total number of authorizations, for *reciprocal* and *deny* accordingly. To not restrict ourselves to a specific scenario, we assume that the only information available is: ($i1$) the number of nodes in the authorization graph G, $|V|$, ($i2$) the number of *deny* edges in G, $|d|$, ($i3$) the number of *reciprocal* edges in G, $|m|$, and ($i4$) the number of *allow* edges in G, $|a|$, such that $|E| = |a| + |m| + |d|$.

Example 12 illustrates how one can compute the probability that the size of the view of a random person s is one, i.e., $P(|View_s| = 1)$.

Example 12. Consider a node s with two incoming and two outgoing edges. s has a view of size 1 if s has either C_1, C_2, C_3 or C_4, where:

[(C_1) One *allow* and one *deny* incoming edge.
(C_2) One *allow* incoming edge, one *reciprocal* incoming edge and one *deny* outgoing edge pointing to the node the *reciprocal* incoming edge originates.
(C_3) One *reciprocal* incoming edge, one *deny* incoming edge, and one *reciprocal* or *allow* outgoing edge pointing to the node the *reciprocal* incoming edge originates.
(C_4) Two *reciprocal* incoming edges, one *reciprocal* or *allow* outgoing edge, and one *deny* outgoing edge such that the outgoing edges point to the nodes the *reciprocal* incoming edges originate.

Then, to compute $P(|View_s| = 1)$, it suffices to sum up the individual probabilities of all the above cases.

Example 12 shows that the distinction between *allow* and *reciprocal* outgoing edges is not relevant in cases C_2, C_3 and C_4. Then, to simplify the computation of $P(|View_s| = N)$ in these cases, we treat *allow* and *reciprocal* outgoing edges as belonging to one group. Lemma 6 proves the correctness of this simplification.

Lemma 6. *Let (1) a value $r \in \mathbb{N}$, (2) a multiset $X = X_1 \cup X_2 \cup X_3$, and (3) a multiset $Y = Y_1 \cup X_3$, where X_1, X_2 and X_3, and Y_1 and X_3 are pairwise disjoint* multisets, their corresponding underlying set is a unit set, and $|Y_1| = |X_1| \cup |X_2|$, be given. Let A_r be the event of choosing r elements from the multiset X such that the chosen elements belong either to the submultiset X_1 or to X_2, and let B_r be the event of choosing r elements from the multiset Y such that the chosen elements belong to the submultiset Y_1. Then

$$P(A_r) = P(B_r) = \frac{\binom{|Y_1|}{r}}{\binom{|Y|}{r}} \tag{5}$$

For readability, we omit the proofs of all lemmas and theorems of this section. All proofs are in our technical report [17].

We now compute the total number of possible graphs (possible outcomes) $|\mathcal{G}|$ that can be built with $|a|$ *allow*, $|m|$ *reciprocal* and $|d|$ *deny* edges. We use $|\mathcal{G}|$ to compute $P(|View_s| = N)$.

Lemma 7. *Given (1) an authorization graph $G = (V, E)$, (2) $|d|$ deny edges, (3) $|m|$ reciprocal edges and (4) $|a|$ allow edges, the number of graphs $|\mathcal{G}|$ that one can build is:*

$$|\mathcal{G}| = \binom{|E|}{|a|} \cdot \binom{|E| - |a|}{|m|} \tag{6}$$

To compute $P(|View_s| = N)$, we now generalize Example 12. To do so, we establish some notions.

Notation: Given an incoming edge of node s coming from u, the *corresponding outgoing edge* is the edge of s pointing to u. Given an outgoing edge of node s pointing to u, the *corresponding incoming edge* is the edge of s coming from u. Given a node s, (1) In_a is the number of *allow* incoming edges of s, (2) In_m is number of *reciprocal* incoming edges, (3) In_d is number of *deny* incoming edges, (4) In_{m1} is number of *reciprocal* incoming edges with a corresponding *allow* or *reciprocal* outgoing edge, (5) In_{m2} is number of *reciprocal* incoming edges with a corresponding *deny* outgoing edge, (6) Out_{am} is number of *allow* or *reciprocal* outgoing edges with a corresponding *reciprocal* incoming edge, (7) Out_d is number of *deny* outgoing edges with a corresponding *reciprocal* incoming edge. Further, given a node s, an integer $N \leq |E_{in}|$, In_a, In_m, In_d, In_{m1}, In_{m2}, Out_{am}, and Out_d, such that $N = In_a + In_{m1}$, s_{In_a} is the event of s having In_a edges. s_{In_m} is the event of s having In_m edges. s_{In_d} is the event of s having In_d edges. $s_{Out_{am}}$ is the event of s having Out_{am} edges. s_{Out_d} is the event of s having Out_d edges. Events s_{In_a}, s_{In_m}, s_{In_d}, $s_{Out_{am}}$ and s_{Out_d} are what we call *dependent events*.

Lemma 8. *Let $|E_{in}|$, In_a, In_m, In_d, In_{m1}, In_{m2}, Out_{am} be given.*

1. *The number of incoming edges is: $|E_{in}| = In_a + In_m + In_d$.*
2. *The number of reciprocal incoming edges is: $In_m = In_{m1} + In_{m2}$.*
3. *The number of reciprocal or allow outgoing edges with corresponding reciprocal incoming edges is $Out_{am} = In_{m1}$.*

4. The number of deny outgoing edges with corresponding reciprocal incoming edges is $Out_d = In_{m2}$.

We now compute the individual probabilities of each case in Example 12 in the general case. The general case is the probability $P(s_E)$ that a node s chosen at random has In_a, In_m, In_d, Out_{am} and Out_d edges. Lemma 9 computes $P(s_E)$, i.e., the joint probability of the dependent events s_{In_a}, s_{In_m}, s_{In_d}, $s_{Out_{am}}$, s_{Out_d}.

Lemma 9. *Let (1) an authorization graph $G = (V, E)$ with $|d|$ deny edges, $|m|$ reciprocal edges, $|a|$ allow edges, (2) an integer $N \leq |E_{in}|$, and the values (3) In_a, In_m, In_d, Out_{am}, and Out_d such that $N = In_a + In_{m1}$ be given. The probability $P(s_E)$ that a random node s has In_a, In_m, In_d, Out_{am} and Out_d edges is:*

$$P(s_E) = P(s_{In_a}) \cdot P(s_{In_m} | s_{In_a}) \cdot P(s_{In_d} | s_{In_a} \cap s_{In_m}) \qquad (7)$$
$$\cdot P(s_{Out_{am}} | s_{In_a} \cap s_{In_m} \cap s_{In_d}) \cdot P(s_{Out_d} | s_{In_a} \cap s_{In_m} \cap s_{In_d} \cap s_{Out_{am}})$$

The formulas of each individual probability together with the corresponding proofs are in our technical report [17].

Theorem 1 computes the probability that a node s chosen at random has a view size equal to a given $N \leq |E_{in}|$, depending on the share of *reciprocal* to *deny* and *allow* authorizations of the population. Theorem 1 sums up all the individual cases of Example 12 in the general scenario. $P(|View_s| = N)$ allow us to study in Sect. 9.2 how the changes in the authorizations, i.e., replacements of *allow* or *deny* authorizations with *reciprocal* ones, affect the size of the view of s.

Theorem 1. *Let an authorization graph $G = (V, E)$, an integer $N \leq |E_{in}|$, $|d|$ deny edges, $|m|$ reciprocal edges and $|a|$ allow edges be given. The probability $P(|View_s| = N)$ that a node s chosen at random has a view of size N is:*

$$P(|View_s|=N) = \sum_{In_{m1}=0}^{min\{N,|m|\}} \sum_{In_{m2}=0}^{min\{|E_{in}|-N,|m|\}} P(s_E) \qquad (8)$$

9 Experiments

This section comprises experimental analyses on the performance impact of *reciprocal* authorizations in making an access decision, the influence of *reciprocal* authorizations on the size of the view of a person, and the experimental validation of QF and FQ complexity analyses.

9.1 Impact of Reciprocal Authorizations

Here, we study the performance when deciding if an access request is authorized.

In Sects. 3 and 5, we showed two different implementations of *reciprocal* authorizations: (1) a basic implementation, which can be used in scenarios with resources with the same sensitivity degree, e.g., LBS, and (2) an extended implementation, trust-based authorizations, designed for scenarios with resources of varying sensitivity, such as health records. Both implementations behave differently regarding performance because to decide if an access request is authorized, the extended implementation also has to evaluate the sensitivity of the resources and the trust value of the users. Hence, our experiments consider scenarios with resources with the same and with different sensitivity.

Next, given an access request $Req = \langle s, \; op, res_u \rangle$, deciding if Req is authorized involves only the authorizations assigned by u to s and the ones assigned by s to u. Therefore, to evaluate the performance on the decision of Req, we alternate the decision-types of the authorizations assigned by u to s and the ones assigned by s to u. Considering this alternation and the possible scenarios regarding resources, we identify the following cases for our experimental analysis:

- *Scenario 1 - Resources with the same sensitivity degree:* Table 2 shows all possible authorizations that users u and s can assign to each other. Based on the assigned authorizations, this scenario has two cases: *S1-Case 1* $\langle Allow/Deny, * \rangle$ and *S1-Case 2* $\langle Reciprocal, * \rangle$. Each case refers to the access request decision of Req considering the authorizations indicated in the column headers and row headers of Table 2. For instance, *S1-Case 1* $\langle Allow/Deny, * \rangle$ refers to the access request decision of Req, given that u has assigned an *allow* or *deny* authorization to s, and s assigns an authorization with any decision-type to u.
- *Scenario 2 - Resources with different sensitivity degree:* Table 3 shows all possible trust-based authorizations that users u and s can assign to each other. Based on these authorizations, this scenario has three cases: *S2-Case 1* $\langle$ *Trust Allow/Deny, * $\rangle$, *S2-Case 2.1* $\langle$ *Trust Reciprocal, Allow/Deny* $\rangle$, and *S2-Case 2.2* $\langle$ *Trust Reciprocal, Reciprocal* $\rangle$. Each case refers to the access request decision of Req considering the trust-based authorizations indicated in the column headers and row headers of Table 3.

Table 2. Experiment Cases - Resources with the same Degree of Sensitivity

s to u	u to s		
	Allow	Deny	Reciprocal
Allow	S1-Case 1		S2-Case 2
Deny	$\langle Allow/Deny, * \rangle$		$\langle Reciprocal, * \rangle$
Reciprocal			

Table 3. Experiment Cases – Resources with Different Degrees of Sensitivity

s to u	u to s		
	Trust Allow	Trust Deny	Trust Reciprocal
Trust Allow	S1-Case 1 $\langle Trust\ Allow/Deny, *\rangle$		S2-Case 2.1 $\langle Trust\ Reciprocal, Allow/Deny\rangle$
Trust Deny			
Trust Reciprocal			S2-Case 2.2 $\langle Trust\ Reciprocal, Reciprocal\rangle$

In Scenario 2, given $Req = \langle s, op, res_u\rangle$, if u has assigned a trust- based *reciprocal* authorization to s, one has to differentiate between S2-Case 2.1 $\langle$ *Trust Reciprocal, Allow/Deny* $\rangle$ and S2-Case 2.2 $\langle Trust\ Reciprocal, Reciprocal\rangle$. In S2-Case 2.2, deciding whether Req is authorized requires an evaluation of the sensitivity of the resources and the trust values of the users involved in the authorizations. This evaluation is not needed if s has assigned a trust- based *allow* or *deny* authorization to u. Regarding *reciprocal* authorizations in S1-Case 2, this differentiation is not needed either because here, the resources have the same sensitivity. Our implementation of *reciprocal* authorizations does not compute the sensitivity of resources and the trust values of users.

Experiment Setup. We use the dataset Epinion, a network dataset of users connected with directed edges. The network has 131828 users and 841372 edges. The network has 30.9% reciprocal edges, i.e., the proportion of edges for which an edge in the opposite direction exists. We use the network dataset to build an authorization graph G as follows: First, the users represent the vertices of G, and the edges of the network represent the labeled directed edges of G. An edge (u, s) indicates that user u has assigned an authorization to s. Second, we label the non-reciprocal edges with labels *allow*, *deny*, or *reciprocal* randomly. The label of an edge (u, s) represents the decision-type of the authorization that u has assigned to s. Third, for each pair of reciprocal edges of the form (u, s) and (s, u), we assign to one of the edges a label *reciprocal*, and to the reciprocal edge a label *allow* or *reciprocal*, selected at random. Next, since we adhere to the *default- deny* principle, we assume that two vertices u, s that are not connected through an edge (u, s) indicate that u has assigned a *deny* authorizations to s. Finally, to evaluate scenario 2, we assign at random to each labeled edge a trust value from the set $T = \{0.1, 0.2, \cdots 1\}$, and to each of the 131828 users a trust value from T. Additionally, we assume that each user owns a resource, and we assign a sensitivity value (from 1 to 5) to each resource at random.

Query Sample: For each of the cases in scenarios 1 and 2, we select a sample of 1000 access requests for the evaluation. This is, for *S1-Case 1* $\langle Allow/Deny, *\rangle$, we select 1000 *allow* or *deny* edges. Then, for each selected edge (u, s), we measure the time needed to decide if $Req = \langle s, op, res_u\rangle$ is authorized. Next, for *S1-Case 2* $\langle$ *Reciprocal, * $\rangle$, we select 1000 *reciprocal* edges and create and evaluate the access request as in the previous case. We repeat the procedure for each case in scenario 2. In total, we run our experiments for 5000 access requests.

Results. Figures 4(a) and (b) show the average time required to decide if an access request is authorized for the cases in scenarios 1 and 2, respectively. The x-axis represents the different cases in each scenario, and the y-axis indicates the average time for the decision.

Discussion: Figure 4(a) illustrates that determining the authorization status in *S1-Case 1* $\langle Allow/Deny, *\rangle$ is faster (approximately 1.6 times) than in *S1-Case 2* $\langle Reciprocal, *\rangle$. In *S1-Case 2*, for an access request $\langle s, op, res_u \rangle$, one must consider authorizations assigned by both u to s and by s to u. Regarding Scenario 2, Fig. 4(b) shows that deciding if an access request is authorized in *S2-Case 1* $\langle Trust\ Allow/Deny, *\rangle$ is faster than in *S2-Case 2.1* $\langle Trust\ Reciprocal, Deny\rangle$ and in *S2-Case 2.2* $\langle Trust\ Reciprocal, Allow/Reciprocal\rangle$. Actually, since we omit the trust evaluation for trust- based *allow* and *deny* authorizations, scenario *S2-Case 1* $\langle Trust\ Allow/Deny, *\rangle$ is the same as *S1-Case 1*. From the *reciprocal* authorizations cases, *S2-Case 2.1* requires less time to decide if an access request is authorized (1.4 faster than in S2-Case 2.2). Namely, S2-Case 2.1 considers that an *allow* or *deny* authorization has been assigned from s to u. Then, trust and sensitivity evaluations for the *allow* and *deny* authorizations become unnecessary. Scenario 1 outperforms Scenario 2, which is unsurprising since Scenario 1 avoids evaluating resource sensitivity and user trust values. This performance gap, however, remains acceptable, providing users with enhanced access control flexibility.

9.2 Experimental Analysis Size of the Views

We run two sets of experiments, named *Reciprocal/Deny* and *Reciprocal/Allow*. Their goal is to analyze how the ratios of *reciprocal* to *deny* and of *reciprocal* to *allow* authorizations, respectively, affect the probability that a random person can read the position of a given number of persons $N \leq |E_{in}|$, i.e., $P(|View_s| = N)$.

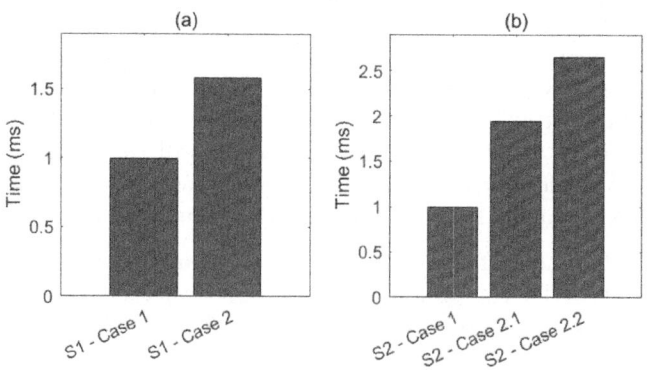

Fig. 4. Access Request Decision – Performance Evaluation

Experiment Setup. We create 100 authorization graphs, 50 graphs for each set of experiments. All graphs have $|V| = 100$ nodes and $|E| = 9900$ edges. We construct the graphs of both experiments by starting with a graph that has 50% *allow* edges, 50% *deny* edges, and 0 *reciprocal* edges. All the graphs fulfill a uniform distribution regarding the number of *allow*, *reciprocal*, and *deny* edges.

For the *Reciprocal/Deny* experiments, we then modify the percentage of *reciprocal* and *deny* labeled edges by increasing the former one in steps of 1 while decreasing the latter at the same rate. The percentage of *allow*-edges remains unchanged. For each setting, we compute $P(|Views| = N)$. We consider three values for N, namely $N = 60$, $N = 80$, $N = |E_{in}|$. For *Reciprocal/Deny* experiments, instead of decreasing the share of *deny*-edges, we decrease the one of *allow*-edges.

Results. Figure 5a for the *Reciprocal/Deny* experiments and Fig. 5c for the *Reciprocal/Allow* experiments show the probability that a node s chosen at random can read the position of N persons, $P(|Views_s| = N)$, contingent on the percentage of *reciprocal* and *deny* authorizations. Figure 5b and Fig. 5d are semi-log plots corresponding to Fig. 5a and Fig. 5c, respectively, which give more emphasis to smaller probabilities.

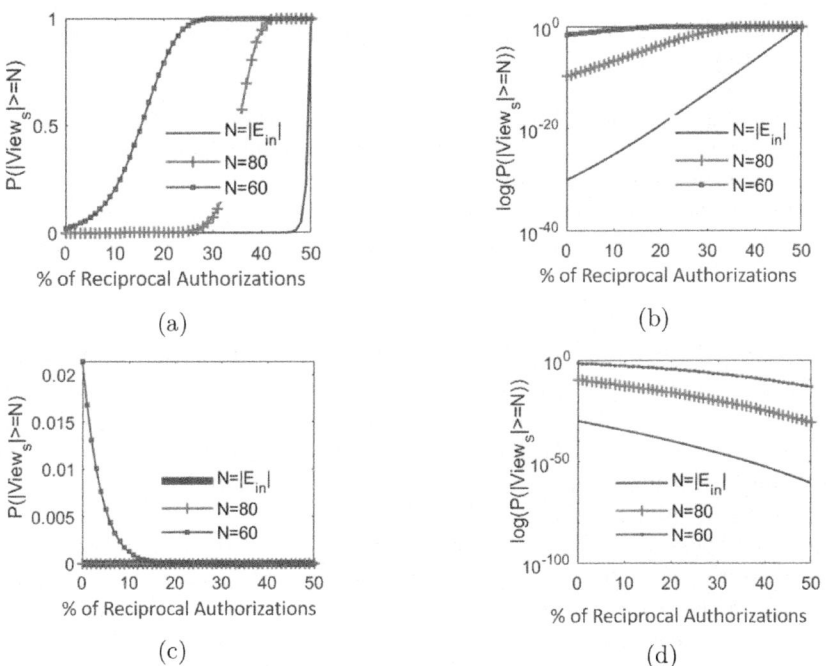

Fig. 5. $P(|Views_s| = N)$ - *Reciprocal/Deny* and *Reciprocal/Allow* Experiments

Discussion: The impact of changing *deny* to *reciprocal* differs depending on the view size. Figure 5b shows that, for larger views, the impact of replacing *deny* authorizations with *reciprocal* ones is higher. For instance, if 5% of *deny* authorizations are replaced by *reciprocal* ones, $P(|View_s| = |E_{in}|)$ increases by a factor of 1000. For $P(|View_s| = 60)$, in turn, the increase is by a factor of 1.3. However, if one is interested in a high probability of having a specific view size, say an 80% chance, the results depend on the target view size. With small values, it is necessary to replace a smaller percentage of *deny* authorizations with *reciprocal* ones to reach this probability. For instance, given 50% of *allow* and 50% of *deny* authorizations, $P(|View_s| = 60)$ is ≈ 0.021. If only 20% of *deny* authorizations are replaced by *reciprocal* ones, this probability increases to 0.8. Consider now $P(|View_s| = |E_{in}|$. Then, if we want this probability to increase to 0.8, one has to replace 49.5% of *deny* authorizations, i.e., almost all *deny* authorizations, with *reciprocal* ones. As the percentage of *reciprocal*-edges increases and the percentage of *allow*-edges decreases in the same proportion, Fig. 5d shows that, for larger views, the impact of replacing *deny* authorizations with *allow* ones is higher. For instance, if 5% of *deny* authorizations are replaced by *reciprocal* ones, $P(|View_s| = |E_{in}|)$ decreases by a factor of 100, whereas $P(|View_s| = 60)$ decreases by a factor of 3.6. Then, the decrease of these probabilities depends on the view size.

9.3 Experimental Validation of the Complexity Analysis of QF and FQ

Our complexity analysis of QF and FQ in Sect. 7 already allows us to compare both approaches. However, since that analysis covers the worst case, experimental results are needed (1) as validation and (2) to determine how far the worst case deviates from the concrete behavior of individual queries. To implement the QF approach, we use the existing R-tree index structure, for spatial querying, from Oracle, and the remaining implementation was done in Java.

Experiment Setup. In Sect. 7, we found, based on the complexity analysis, that the parameters that affect the performance of FQ and QF are (1) the number of persons n, (2) the size of the view $|View_s|$ of a given person s, (3) the parameter k of the kNN query and (4) the value $\delta = k_{real} - k$. Similarly to the complexity analysis, for simplicity, we set δ to 0 and k to 20. We set the remaining parameters, n and $|View_s|$, as follows:

Number of persons n: We create a dataset with 317080 persons, the size of the DBLP dataset. To assign a position to each person, we choose a random physical position from the Tokyo dataset [23], which contains 573703 real check-ins.

Size of the view of a person s, $|View_s|$, and query sample: We chose 1500 persons at random from the 317080 persons and assigned the authorizations so that we have 15 classes of different-sized views (from 50 to 40000). For each class, we have 100 persons with the respective view size, i.e., 1500 queries in 15 different classes in total.

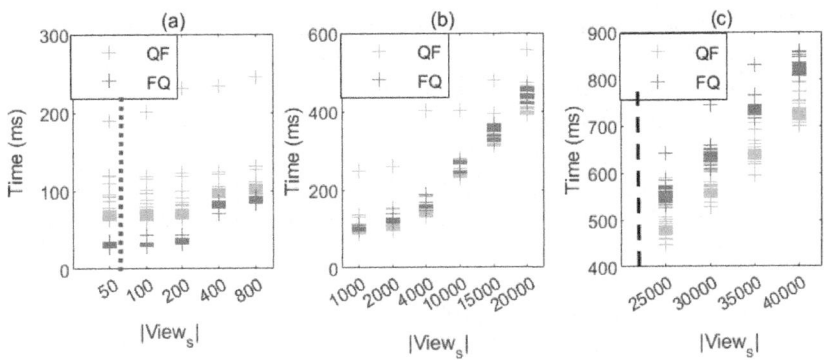

Fig. 6. Comparison of the QF and FQ Approaches for kNN Queries

Experiment Results. Figure 6 shows a comparison of the query-processing times for kNN queries with QF and FQ. We have grouped the persons of our query sample by the size of their view and have plotted the query-processing time. We exclude the database-connection time and network-communication time from the run time reported. The dotted line is the average size of the view in DBLP, i.e., 64.98. The dashed line in Fig. 6(c) is the size of the view for which the performance of both approaches is equal, i.e., $View_{eq} \approx 23032,3$.

Discussion. For real scenarios, with a view size equal to 64.98, FQ performs better than QF for all queries. These results are in line with our complexity analysis, and one may interpret them as an indication that our analysis also holds for the average case. Next, these findings remain correct for a view size up to 800, which is higher than the largest average view size in real scenarios, i.e., 520 (Sect. 7). However, as Fig. 6(b) shows, with a view size between 1000 to 20000, the processing times of most of the queries with QF are lower than that of the ones with FQ, in contrast to our complexity analysis, which can be expected since our analysis has focused on the worst case. Figure 6(c) shows that the processing times of most of the queries with QF, or all the queries in the case of the last two groups, i.e., 35000 and 40000, are lower than that of the ones with FQ. These results indicate that for a view size greater than the value $View_{eq}$ (dashed line), our complexity analysis holds even for the average case.

10 Related Work

In line with our work, the authors in [24] introduce MuAC, a policy language that allows users to state the conditions under which they are open to exchange their resources and what they require in return. The authors in [25] define a low-level language for capturing access requests in collaborative environments, addressing infinite or reusable assets. However, these approaches overlook conflicts like revocation fraud and resource sensitivity. In [26], an analysis of existing language policies for trust negotiation and resource exchange is presented, with

PlexC [27] identified as the sole policy language handling trust negotiation but lacking the benefits of exchanging resources reciprocally. Trust-based authorizations, integrating trust and reputation scores, are discussed in [28–31]. Putra et al. [28] investigate blockchain implementation for IoT devices, while Singh et al. [29] incorporate various trust types such as authentication and privacy. Zahoor et al. [30] model temporal and trust aspects into authorizations, and Dimitrakos et al. [31] address trust aspects with continuous monitoring and policy re-evaluation. However, these approaches do not consider reciprocal benefits between users.

Several IoT studies explore reciprocity, with a focus on either mutual authentication [32] or reciprocal spectrum sharing [33]. Mutual authentication ensures secure communication by authenticating both IoT devices and servers before allowing data exchange. Reciprocal spectrum sharing facilitates the coexistence of multiple wireless communication technologies in the same frequency bands, optimizing resource utilization. Next, reciprocity-based incentive mechanisms [8,34,35] are commonly employed to combat free-riders in mobile peer-to-peer systems (MP2P), where the majority of peers exhibit selfish behavior and are reluctant to share their limited resources. However, these works primarily investigates spectrum-sharing dynamics, efficiency and dynamics of the MP2P system rather than formalizing access requests.

Besides access control models, encryption techniques are explored for data confidentiality [36–38]. In [36], resources are encrypted with keys tied to access authorizations, and users receive decryption keys accordingly. In [37], the authors encrypt data with an access structure, necessitating specific attribute values for decryption keys. The authors in [38] propose an encryption approach for LBS data. However, these methods do not address reciprocal access considerations. However, these approaches do not incorporate reciprocal access considerations. Other research formalizes and verifies authorization constraints in RBAC and its extensions using techniques like Colored Petri-Nets [39] and the Unified Modeling Language UML [40]. These efforts aim to ensure authorization design and consistency through model checking and graphical representations but are restricted to *allow* and *deny* authorizations.

11 Conclusions

Reciprocity is a powerful determinant of human behavior. However, none of the existing access control models explicitly supports it. In this paper, we have proposed a new type of authorization called *reciprocal*. It allows users to grant access to their resources to users that permit them the same. We have extended ABAC to incorporate *reciprocal* authorizations and have formally defined their syntax and semantics. We consider scenarios where resources have the same and different sensitivity, i.e., general case. Our generalization lets owners assign the degree of sensitivity of their resources themselves, but their peers can evaluate such an assignment a posteriori. Based on the evaluation by others, each user receives a trust value. When using *reciprocal* authorizations, users can then express the minimum level of trust their peers should have to exchange resources with them. We call this extension trust- based authorizations. Next, we have conducted an investigation into the implications of *reciprocal* authorizations on service outcomes, particularly in the context of their deployment within Location-Based Services (LBSs). Two distinct approaches were proposed and subjected to rigorous analysis utilizing complexity metrics, which provided insights into their respective performance characteristics. Furthermore, experimental validation was conducted to corroborate our analytical findings. Additionally, we have evaluated the influence of *reciprocal* authorizations on access decision performance, as well as examined the effects of varying *reciprocal* to *deny* and *allow* authorization ratios on the size of the view of a given user.

In the future, we plan to consider expanding upon the design space for both *reciprocal* authorization and *trust-based* authorization We believe that a more comprehensive exploration of alternative designs and approaches would contribute to a deeper understanding of their implications and potential applications in practical settings. Moving forward, investigating the application of encryption techniques to achieve data secrecy and confidentiality, as well as reciprocally generating and distributing encryption-decryption keys, poses an intriguing avenue for exploration. Another direction is to extend *reciprocal* authorizations by considering different resources with more than one controller, i.e., users who can regulate access to the resource and users who own more than one resource with different degrees of sensitivity. Next, studying settings where resources have multiple owners, i.e., multiparty access control models, remains future work. These kinds of access control models place interesting challenges aligned with consensus-reaching processes where reciprocity may play a significant role.

Acknowledgements. This work was partially funded by the German Research Foundation (DFG) as part of the research Datenschutzkonforme Verwaltung relationaler Datenbestände (DFG; ref. nb BO 2129/13-1).

References

1. Hu, V.C., Ferraiolo, D., Kuhn, R., Friedman, A.R., et al.: Guide to attribute based access control (ABAC). NIST Special Publication **800**(162) (2013)
2. Fehr, E., Fischbacher, U., Gächter, S.: Strong reciprocity, human cooperation, and the enforcement of social norms. Hum. Nat. **13**(1) (2002)
3. Falk, A., Fischbacher, U.: A theory of reciprocity. Games Econ. Behav. **54**(2) (2006)
4. Sandhu, R.S., Coynek, E.J., Feinsteink, H.L., Youmank, C.E.: Role-based access control models. IEEE Computer **29**(2) (1996)
5. Oh, S., Park, S.: Task–role-based access control model. Inf. Syst. **28**(6) (2003)
6. Thion, R., Lesueur, F., Talbi, M.: Tuple-based access control: a provenance-based information flow control. In: Proceedings of the 30th SAC. ACM (2015)
7. Atluri, V., Shin, H., Vaidya, J.: Efficient security policy enforcement for the mobile environment. J. Comput. Secur. **16**(4) (2008)
8. Zhou, R., Hwang, K.: Powertrust: a robust and scalable reputation system for trusted peer-to-peer computing. IEEE TPDS **18**(4), 460–473 (2007)

9. Hutter, C., Lorch, R., Bohm, K.: Evolving cooperation through reciprocity using a centrality-based reputation system. In: 2011 IEEE/WIC/ACM WI-IAT, vol. 2, pp. 264–271. IEEE (2011)
10. Suntaxi, G., El Ghazi, A., Böhm, K.: Mutual authorizations: semantics and integration issues. In: Proceedings of the 24th ACM SACMAT, SACMAT 2019, Toronto, Canada. ACM (2019)
11. Hu, H., Ahn, G.-J.: Multiparty authorization framework for data sharing in online social networks. In: Li, Y. (ed.) DBSec 2011. LNCS, vol. 6818, pp. 29–43. Springer, Heidelberg (2011). https://doi.org/10.1007/978-3-642-22348-8_5
12. Bertino, E., Bonatti, P.A., Ferrari, E.: TRBAC: a temporal role-based access control model. ACM TISSEC **4**(3) (2001)
13. Pagnia, H., Gärtner, F.: On the impossibility of fair exchange without a trusted third party. Technical report, TUD-BS-1999-02, Darmstadt University (1999)
14. Eckey, L., Faust, S., Schlosser, B.: Optiswap: fast optimistic fair exchange. In: ASIA CCS 2020, New York, USA (2020)
15. Dziembowski, S., Eckey, L., Faust, S.: Fairswap: how to fairly exchange digital goods. In: Proceedings of the 2018 ACM SIGSAC, pp. 967–984 (2018)
16. Shi, J., Zhu, H., Ge, F., Jiang, T.: On the soundness property for SQL queries of fine-grained access control in DBMSs. In: Eighth IEEE ICIS. IEEE (2009)
17. Suntaxi, G., El Ghazi, A., Böhm, K.: On mutual authorizations: semantics, integration issues, and performance (2019). https://publikationen.bibliothek.kit.edu/1000093936
18. Robusto, C.C.: The cosine-haversine formula. Am. Math. Mon. **64**(1), 38–40 (1957)
19. Mehta, D.P., Sahni, S.: Handbook of Data Structures and Applications, 2nd edn. Chapman and Hall/CRC (2018)
20. Mahapatra, R.P., Chakraborty, P.S.: Comparative analysis of nearest neighbor query processing techniques. Procedia Comput. Sci. **57**, 1289–1298 (2015)
21. Yi, X., Paulet, R., Bertino, E., Varadharajan, V.: Practical k nearest neighbor queries with location privacy. In: 2014 IEEE 30th ICDE. IEEE (2014)
22. Lattanzi, S., Singer, Y.: The power of random neighbors in social networks. In: Proceedings of the Eighth ACM International WSDM (2015)
23. Yang, D., Zhang, D., Zheng, V., Yu, Z.: Modeling user activity preference by leveraging user spatial temporal characteristics in LBSNs. IEEE Trans. SMC **45**(1), 129–142 (2015)
24. Ceragioli, L., Degano, P., Galletta, L.: MuAC: access control language for mutual benefits. In: ITASEC, pp. 119–127 (2020)
25. Ceragioli, L., Degano, P., Galletta, L., Basin, D., Pugliese, R.: Access control policies across abstraction layers (2022)
26. Kolar, M., Fernandez-Gago, C., Lopez, J.: Policy languages and their suitability for trust negotiation. In: Kerschbaum, F., Paraboschi, S. (eds.) DBSec 2018. LNCS, vol. 10980, pp. 69–84. Springer, Cham (2018). https://doi.org/10.1007/978-3-319-95729-6_5
27. Le Gall, Y.G., Lee, A.J., Kapadia, A.: Plexc: a policy language for exposure control. In: Proceedings of the 17th ACM SACMAT, pp. 219–228 (2012)
28. Putra, G.D., Dedeoglu, V., Kanhere, S.S., Jurdak, R., Ignjatovic, A.: Trust-based blockchain authorization for IoT. IEEE Trans. Netw. Serv. Manag. **18**(2), 1646–1658 (2021)
29. Singh, S.: Trust based authorization framework for grid services. J. Emerg. Trends Comput. Inf. Sci. **2**(3) (2010)

30. Zahoor, E., Perrin, O., Bouchami, A.: Catt: a cloud based authorization framework with trust and temporal aspects. In: 10th IEEE CollaborateCom, pp. 285–294. IEEE (2014)
31. Dimitrakos, T., et al.: Trust aware continuous authorization for zero trust in consumer IoT. In: 19th IEEE TrustCom, pp. 1801–1812. IEEE (2020)
32. Ma, Q., Tan, H., Zhou, T.: Mutual authentication scheme for smart devices in IoT-enabled smart home systems. Comput. Stand. Interfaces **86**, 103743 (2023)
33. Thakare, S., Patil, A., Siddiqui, A.: The internet of things-emerging technologies, challenges and applications. Int. J. Comput. Appl. **149**(10), 21–25 (2016)
34. Kun, L., Wang, S., Cui, G., Li, M., Bin-Liaqat, H.: Multi-reciprocity policies co-evolution based incentive evaluating framework for mobile P2P systems. IEEE Access **5**, 3313–3321 (2016)
35. Zhu, H., Ding, H., Zhao, Q.-Y., Yan-Ping, X., Jin, X., Wang, Z.: Reputation-based adjustment of fitness promotes the cooperation under heterogeneous strategy updating rules. Phys. Lett. A **384**(34), 126882 (2020)
36. Vimercati, S., Foresti, S., Jajodia, S., Paraboschi, S., Samarati, P.: Encryption policies for regulating access to outsourced data. ACM TODS **35**(2) (2010)
37. Bethencourt, J., Sahai, A., Waters, B.: Ciphertext-policy attribute-based encryption. In: IEEE Security and Privacy, 2007 (2007)
38. Suntaxi, G., Ghazi, A.A.E., Böhm, K.: Preserving secrecy in mobile social networks. ACM TCPS **5**(1), 1–29 (2020)
39. Shafiq, B., Masood, A., Joshi, J., Ghafoor, A.: A role-based access control policy verification framework for real-time systems (2005)
40. Ray, I., Li, N., France, R., Kim, D.: Using UML to visualize role-based access control constraints. In: Proceedings of the 9th ACM SACMAT. ACM (2004)

Device Forensics in Smart Homes: Insights on Advances, Challenges and Future Directions

Sabrina Friedl[✉][iD] and Günther Pernul[iD]

University of Regensburg, Universitätsstraße 31, 93053 Regensburg, Germany
{sabrina.friedl,guenther.pernul}@ur.de

Abstract. Attackers increasingly use smart home Internet of Things (IoT) devices for criminal activities, such as creating botnets, spying, robbery, and theft. Rapid countermeasures and support for digital investigations in the smart home area are essential. Our contribution provides an overview of the types of devices currently being researched, including (1) Smart Speakers and Voice Assistants, (2) Wearables, and (3) Smart Things. Preparatory measures have been identified, such as knowledge of potential digital evidence (PDE) sources and evidence data categories, user(s), device(s), and communication. The paper also outlines extraction approaches for IoT devices in smart homes, including manual, logical, and physical methods. These findings serve as a basis for discussing and comparing current advances in IoT device forensics by category, leading to the extraction of a taxonomy, challenges, and open future work directions. As the use of smart home IoT devices continues to expand, it is crucial to systematize knowledge to enhance the quality of forensic investigations proactively. This paper offers valuable insights into device forensics and provides a roadmap for future research.

Keywords: Internet of Things (IoT) · Device Forensics · Digital Forensics (DF) · Smart Home · Consumer IoT · Evidence Extraction

1 Introduction

The term IoT designates an environment of embedded interconnected computing devices that allows for implementing technologies such as Machine-to-Machine (M2M) communication and context-aware computing [57]. More than 12 billion IoT connections are expected to grow significantly (about 20% p.a. during the next five years) [27]. The high amount of involved IoT devices causes new avenues for attackers to exploit compared to non-digital environments. Attacks on intelligent devices related to personal health, privacy, and safety like pacemakers, infant monitoring systems, or smart locks threaten to turn digital security and privacy threats into physical ones [57].

An essential part of the IoT landscape is the area of smart home IoT, with an estimated industry revenue of 115,7 Billion US$ in 2022 [56]. Smart Homes are simple forms of IoT environments consisting of IoT devices, home and hub gateways, the mobile devices on which IoT apps are installed, and the corresponding cloud servers. Several devices can be controlled through a central device, helping the owner to automate personal tasks, manipulate and monitor devices remotely, or regulate energy consumption [32]. The amount of processed data, often containing sensitive information about people's private behavior, the ability to remote control devices for devious purposes, and an apparent lack of security awareness and scrutiny, make smart home IoT systems an attractive target for cyber-attacks and, consequently an essential subject for security research [16].

To combat vulnerabilities in IoT Systems in general, extensive research has already been conducted regarding how to ensure the safety of IoT systems. An essential subset of security is the reaction to incidents and forensics. Digital Forensics (DF) is concerned with uncovering and interpreting electronic data. The steps associated with this aim comprise a simplified version for IoT (1) Initialization, (2) Acquisition, and (3) Investigation [29]. As the acquired evidence must be suitable to be presented in a court of law, they have to be obtained legally and fulfill the principles of forensic soundness, meaning that the forensic process and extraction of evidence follow the standards for legal admissibility and obey rules that protect extracted evidence from damage and manipulation. IoT devices can be investigated regarding their involvement in incidents as tools, targets, or witnesses [43]. In IoT forensics (IoTF), in particular, the different levels of sources where data can be acquired, such as sensors on a (1) Device Level, internal/external networks on a (2) Network Level, or on a (3) Cloud Level. These three levels are often used to designate subdivisions in the field. Unlike traditional DF, evidence may be extracted from computers or network devices and a diverse range of IoT devices and their related network connections and cloud servers [57]. The challenges that arise from this fact support the need for research in IoT device forensics, thus raising the following research questions and providing the respective contribution.

RQ1: *What IoT device types are investigated in smart homes?*
RQ2: *How and what evidence is extracted from IoT devices?*
RQ3: *How advanced device forensics in smart homes?*

The remainder of the paper is structured as follows. First, related work is shortly reviewed in Sect. 2.1 and delimited. Then, the method and scope are set in Sect. 2.2. The extracted research on device forensics in smart homes is then analyzed according to the identified categories and systematized by extracted aspects in Sect. 3. After setting a systematic approach, we discuss advances and compare the approaches in Sect. 4. Here, Sect. 4.1 starts with Smart Speakers and Voice Assistants, followed by Sect. 4.2 Wearables and Sect. 4.3 Smart Things. We provide a novel view with a taxonomy based on the extracted knowledge. Then, Sect. 5 discusses challenges and open research. Finally, we conclude our work and gained insights in Sect. 6.

2 Related Work, Methodology, and Scope

This section lays out related works, the methodological approach for extracting relevant publications regarding the topic device forensics in smart homes. Further, is the scope of the presented work delimited and defined.

2.1 Related Work

There exist various research works [5,33,46,51,62] on IoTF providing reviews, overviews, surveys, and state-of-the-art, mainly including identified challenges, advances, and open future directions. The related work on IoTF and specifically of interest for us smart home IoTF is visualized in Table 1. The table provides insights into similarly structured research on IoTF. While some of them do not consider smart home devices [46,51], others already mention and shortly discuss them as an essential part of IoTF [33,62].

Table 1. Related work on IoTF with a focus on the smart home area.

Source	Year	DF	IoT	SH	Short Description
Oriwoh et al. [51]	2013	✓	✓	×	They suggest a high-level incident response strategy for approaching IoT-based crime scenarios.
MacDermott et al. [46]	2018	✓	✓	×	They discuss the Internet of Anything (IoA), provide forensics evidence handling methods and main challenges in IoTF.
Yaqoob et al. [62]	2019	✓	✓	~	They explore IoT's novel factors affecting DF, create a taxonomy, provide key requirements and open research on IoTF.
Janarthanan et al. [33]	2021	✓	✓	~	They present IoTF challenges of smart home investigations and compare existing frameworks for IoTF in general.
Kaushik et al. [37]	2023	✓	✓	✓	They create a general overview of IoTF for smart homes and discuss how important smart home devices are for analyzing cybercrime.

Abbreviations: DF = Digital Forensics; IoT = Internet of Things; SH = Smart Home
Symbols: ✓ included; ~ partly included; × not included

With the focus area on smart home IoTF or device forensics, we could only identify one recent publication. *Kaushik et al.* [37] provide a very high-level and superficial view of smart home IoTF. They identify various challenges, like heterogeneous IoT devices, a lack of standards and training, data protection and privacy laws, as well as problems with decryption and data corruption. These challenges, approaches, and future directions are provided, but not in a very detailed manner. Compared to this work, we focus on currently applied approaches and techniques and extracted evidence data on the IoT device level in the smart home environment. We provide detailed insights into this young research area. In doing so, we can extract a novel state-of-the-art overview of smart home IoT device forensics, revealing a taxonomy, challenges, and future research opportunities based on three identified device categories.

2.2 Methodology and Scoping

For analyzing the existing forensic approaches concerning smart home IoT devices discussed in the literature, we first conducted a systematic literature research (SLR) based on the principles of *Okoli and Schabram* [50]. As the range of potentially pertinent devices is rather diverse and the list of used terms developed over the research process, we added devices to the corpus for which we found forensic techniques applied. The search term applied and associated settings can be found in Table 2.

Table 2. Settings of SLR.

Purpose	Structure knowledge on smart home device forensics (RQ1-3)
Audience	IoT Forensics Researchers (novice to expert)
Databases	Springer, IEEE Xplore, ACM Digital Library, and ScienceDirect
Keywords	Internet of Things, IoT, wireless, forensics, smart home, smart things, smart device, wearables
Search term	["forensic*" AND ("device" OR "smart home" OR "smart watch" OR "Internet of Things" OR "IoT" OR "smart speaker" OR "wearables" OR "smart vacuum cleaner" OR "smart camera" OR "smart lock")]
Search setting	Metadata (e.g., title, abstract, keywords)

As the field of smart home device forensics is very dynamic and quickly developing, we decided only to include papers published from the beginning of the year 2015 up until the present (January 2024). When conducting the structured literature search with the pre-defined adjustments, 567 results were produced. These then could be reduced to 44 relevant papers by several steps of scanning and sorting out, like title check, abstract reading, and full-text inspection [50].

> **RQ1: What IoT device types are investigated in smart homes?**
>
> The 44 found papers can be put into three main categories, the first being device forensics research related to *Smart Speakers and Voice Assistants*, with 11 results. *Wearables*, a classification that includes devices like fitness trackers and smartwatches, is the second category with 16 identified papers. In contrast, the *Smart Things* category with 17 publications includes devices making a house intelligent (e.g., lights, vacuum robots, cameras, and other sensors). It is not limited to one specific kind of device. The categorization represents the distinct characteristics of the associated devices.

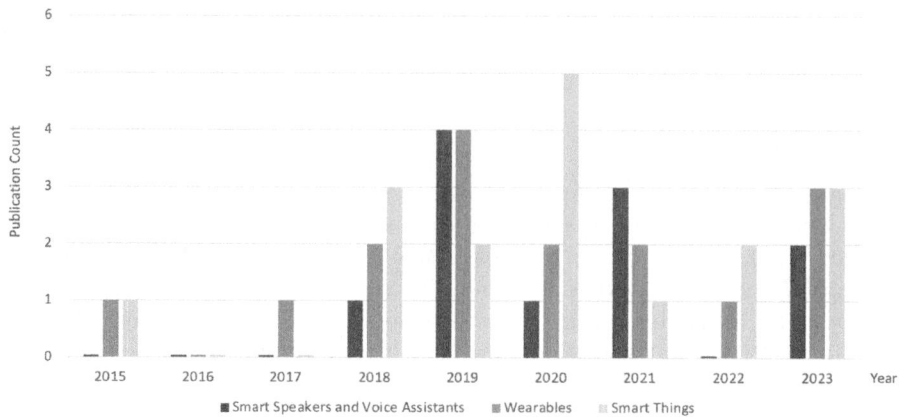

Fig. 1. Relevant publications divided by extracted domains: smart speakers and voice assistants, wearables and smart things.

3 Systematization

This section provides the basis to systematize the extracted research on device forensics in smart homes. Therefore, we start by delineating the two main domains in the IoT. The underlying IoT forensic investigation process follows this. Building on these process steps, we identify PDE, which is accompanied by data acquisition and, thus, applied data extraction approaches for IoT devices. This way, valuable information could be extracted from literature, and, at the same time, the basis could be built for discussing the advances from a structured perspective, providing novel insights.

3.1 Differences in Consumer and Industrial IoT

The IoT encompasses a wide range of connected devices and applications used in private and industrial sectors. However, there are critical differences between consumer and industrial IoT systems in scale, connectivity, security, analytics, and maintenance [3].

Consumer IoT refers to the integration of IoT technology into consumer devices and applications. It encompasses physical devices (e.g., refrigerators, TVs, thermostats, cameras, smart lamps) that are connected to the internet and use sensors to collect, process and share data. Such systems enable remote monitoring and controlling. In difference to Industrial IoT, the consumer IoT aims to provide solutions and convenience at an individual level. These solutions include efficient tracking, improved connectivity, better insights, more control and more convenience. All of these features can be applied to various aspects of daily life, such as entertainment, home security, healthcare and asset tracking. While the smartphone can be seen as the ultimate IoT device for consumers, there are other well-known examples. These include smartwatches, smart glasses, trackers and smart home technologies such as voice or text-controlled home appliances [60].

Industrial IoT describes the integration of IoT technology within an industrial setting, mainly supporting manufacturing. For instance, smart factories utilize IoT capabilities for predictive maintenance, robotics, supply chain insights, quality improvement, optimized automation and smart grids. The industrial IoT also makes use, just like the consumer IoT of the billions of physical devices that are connected to the internet and collect, process and share data. The difference is, that industrial IoT networks are mostly designed as isolated environments, which can positively effect the security of the system. These networked, mission-critical architectures enables the real-time exchange of industrial system information in real time, enabling better situational awareness, better control of system processes and an increased productivity and efficiency [60].

IoT devices, regardless of their application or operating domain, are responsible for monitoring, controlling and improving system connectivity and performance. In terms of basic security objectives, the IoT device must ensure the confidentiality of collected measurements during operation measurements, ensure the information integrity of the data stored or in transit and allow access only to authorized users/parties [48]. However, in comparison generally a clear distinction can be made between consumer and industrial IoT, as they serve different goals and target audiences. Consumer IoT focuses on enhancing and automating individuals' daily lives, while industrial IoT primarily aims to enhance production and maintenance in corporate settings. Additionally, consumer IoT networks are rather small in comparison to industrial IoT networks. These differences affect security, maintenance and analytics of the respective IoT domain [3]. This effect expands to DF and the need for different approaches to investigate certain IoT domains forensically. We categorize our work within the consumer IoT, as private individuals predominantly use smart home devices.

3.2 IoT Forensic Investigation Process

IoTF is becoming increasingly crucial for forensic investigations, as IoT devices can help prove the guilt or innocence of suspects. However, for IoTF to be practical, we must establish an investigative process. As described in the introduction, IoTF is divided into three sub-categories: (1) Cloud forensics, (2) Network forensics, and (3) Device forensics [57]. Not all three categories may be required depending on the nature of the incident investigated. The three steps of an IoT forensic investigation process are shortly detailed following [30].

1. **Initialization.** Preparatory steps are taken before interacting with equipment at the accident scene. This could mean understanding how the respective IoT ecosystem works and thus identifying PDE sources (e.g., crime scene layout).
2. **Acquisition.** IoT devices and data sources are collected in a forensically sound manner. Data to be collected during this step should be preserved, and forensically verified methods and tools used to acquire data.

3. **Investigation.** The investigator reviews the collected data for relevance to the incident and analyses all relevant data to determine the six W's: What happened? Who was responsible? When exactly did the incident happen? Why did the suspect do or not do it? Moreover, How was it done? Once these questions are answered, the investigator concludes the investigation in a report [30].

3.3 Smart Home Digital Evidence

When we start to look into existing approaches on device forensics for IoT-based homes, from a forensics perspective, in a first step, we need to know where PDE can be found in comparison to what is currently researched in IoTF.

Potential Digital Evidence (PDE) Sources and Data. The available sources, where forensic investigators can start to look for evidence, can comprise all technical devices. These may contain, on a device-level, IoT devices (e.g., smart plugs, sockets, doorbells, speakers, voice assistants, fridges, coffee machines, robot vacuum cleaners or shutters) and wearables (smartphones, watches, or clothes) directly connected with a person. For network and cloud-level, these differ. This work focuses on device-level forensics, especially IoT devices in intelligent home environments (cf. 1. Initialization and 2. Acquisition).

In a smart home, each device provides an independent service. At the same time, the associated apps, usually installed on smartphones or IoT hubs, can be used to operate and monitor the device(s). In addition to smartphones, smart speakers such as Google Nest and Amazon Echo can be used as control instances via voice commands. Data from smart speakers, fitness wearables, pacemakers, and biometric devices have previously been used as evidence in court [40]. For example, a file recorded by a smart speaker (Amazon Echo) played a crucial role in proving the innocence of a murder suspect in 2015 [14]. In the same year, recordings from a fitness tracker (Fitbit) were used to prove that a suspect's testimony was false [15]. In 2017, data from a pacemaker was used as evidence in an insurance fraud case [14]. To extract evidence data successfully and thus support various investigations, it is necessary to know what categories of data exist and can be gathered from IoT devices (cf. 3. Investigation).

Based on the literature analysis, we could extract three categories. The category *Device(s)*, represents information about the device itself (e.g., model, firmware, cache, memory image). The next category is the *User(s)*, which contains all information of IoT device owner(s) and user(s) (e.g., to-do/shopping lists, personal and location data, login credentials, browser history, images). The third category *Communication*, maps all information that is exchanged in some way between user(s), device(s), network(s), and cloud present in connection with the investigated IoT devices.

3.4 Extraction Approaches for IoT Devices

Diverse data extraction approaches exist for data from mobile devices. These methods are already used in mobile phone forensics and could be adopted similarly for IoT devices. A general distinction is made into three categories: (1) manual, (2) logical, and (3) physical extraction (cf. 2. Acquisition).

(1) Manual Extraction means when the investigator is confronted with the user interface of a mobile device (e.g., phone, smart speaker, smart TV, IoT hub), scrolling through a normally functioning device and documenting (e.g., screen recording) what the investigator sees. Thus, data can be uncovered by simply clicking through the call log, reading text messages, looking at photos, scanning browsing history [21].

(2) Logical Extraction provides a more in-depth analysis of the data on a device and is the first method of 'extracting' data to examine it separately from the device. This method involves establishing a wired or wireless connection between the device and a forensic workstation to access the device's file system and retrieve copies of the backed-up data [21].

(3) Physical Extraction, also known as physical memory dump, is a technique to capture all data from flash memory chips on mobile devices. It is the most comprehensive but also the most time and resource-consuming method. This is not a copy as in logical capture but a deep dive into the device's file system, giving investigators access to the contents, including deleted or hidden data. This is why many investigators prefer this method - simply because of the possibility of getting a comprehensive picture of the device. Initially, the data received is in raw format and cannot be read. Later, tools are applied to convert this data into a human-readable form. There are three standard techniques for physical data acquisition [21].

- **Chip-On** methods (non-invasive) are Hex Dumping, JTAG (Joint Test Action Group), or ISP (In System Programming). *JTAG* is a non-destructive method manufacturers use to debug their devices before launch. As forensic examiners, we can find these ports and use them to communicate with the processor, which communicates with the memory card to get a complete physical image of the device (byte-by-byte dump) [21]. Like JTAG, *Hex Dumping* is a method of physically extracting raw data stored in flash memory. This involves connecting the forensic workstation to the device and tunneling unsigned code or a bootloader into the device, each of which transmits instructions to dump the memory from the phone to the computer [21]. *ISP* extractions are used when JTAG or Test Access Points (TAPs) are unavailable, and the tester needs to solder a connection directly to the PCB. The difficult part is finding the pinout of the desired device, which shows which pins to solder. This method is usually more difficult because the pins are much smaller than JTAG TAPs, which require a microscope, a much finer soldering tip, and a steady hand. This method also works for passcode-enabled devices but not for encrypted devices [21].

- **Chip-Off** is a destructive technique (invasive) that removes the flash memory chip from the printed circuit board (PCB). This requires cutting the PCB and grinding the PCB to expose the chip's contacts. Once the chip is prepared, the memory registers of the chip are set using the correct adapter and by running a programming application to extract raw data [21].
- **Micro Read** methods involve using a high-powered microscope to view the physical state of the gates. These methods are the most invasive, challenging, technically complex, expensive, and time-consuming. While micro-reading techniques can enable the examiner to extract critical materials and analyze hidden security mechanisms of components on the target device, it remains an arduous task [21].

> RQ2: How and what evidence is extracted from smart homes?
>
> In the first step, the knowledge of potential PDE sources helps to map them. In the second step, these sources are collected and analyzed with one or more suitable extraction methods: manual, logical, and physical. In the third step, PDE is located and forensically sound extracted with the help of evidence categories within smart homes: device(s), user(s), and communication.

4 IoT Device Forensics Advances in Smart Homes

The following Sects. 4.1, 4.2, and 4.3 briefly discuss the advances of device-level forensic approaches in smart home environments from 2015 until today (2024). The advances are discussed according to the categories as defined in Sect. 2.2, Smart Speakers and Voice Assistants (Sect. 4.1, Table 3), Wearables (Sect. 4.2, Table 4), and Smart Things (Sect. 4.3, Table 5). A quick view of the associated tables reveals the extracted data, applied extraction approach, used tools, and investigated devices, visualizing the evolution of the respective category. Additionally, we provide an overview and structure the knowledge on the state-of-the-art research, by developing a literature-based taxonomy, as seen in Fig. 2. The generic visualization represents the actual researched areas in the literature, so research gaps can be visible and an overview of existing findings and challenges can be extracted. This allowed us to create an overview of smart home IoT device forensics partitioned into three areas: first, the evidence types extracted; second, the device types investigated; and third, the evidence extraction method applied on IoT devices.

4.1 Advances on Smart Speakers and Voice Assistants

The smart speakers and voice assistants category is dominated by research on products from the Amazon Echo line and Amazon's proprietary voice assistant, Alexa. Thus, related findings and advances are discussed hereafter and visualized in Table 3.

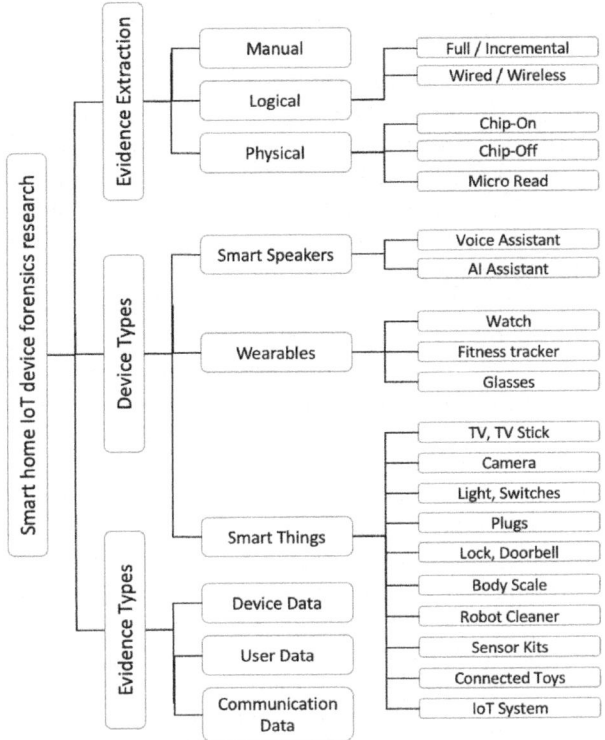

Fig. 2. Literature-based taxonomy on smart home IoT device forensics, with evidence extraction methods, device and evidence types.

Smart Speakers and Voice Assistants. The first category extracted deals with smart speakers and associated voice assistants, where earlier research is mainly focused on data extraction from the Amazon server APIs Alexa uses, an approach pioneered by *Chung et al.* [13], who are cited by all following papers relying on APIs for data extraction. *Orr and Sanchez* [52] examine what data can be extracted manually from APIs through Alexa's web-based interface, the mobile companion app, and a logical backup of its workstation, and how the obtained data might be helpful in a police investigation. Factual information on the content and value of the cache data was not provided, as it would have required jailbreaking the phone. *Yildirim et al.* [63] work with the Amazon Echo Plus 2 and the Google Home Mini. They create fake activities for the examination, such as changing device names, making artificial routines, and custom skill development. As a result of the investigations, they provided information on which kind of PDE can be found in the assistant's activities. They mainly extract user-related data manually without applying any additional tools. *Li et al.* [43] additionally extract firmware data with the help of an Alexa and a Raspberry Pi simulating an Amazon Echo by using similar firmware. However, there is no proof that this method can be directly applied to a standard Echo device. They present four categories to their extracted data from the echo environment, which can be attributed to Device, Connectivity, User, and Application Data. *Pawlaszczyk et al.* [53] focus on the examination of the Echo Dot 3. To do so, they did a chip-off and then performed a hardware-side analysis on the chip to get a firmware dump, revealing OS and file system structures. Further, based on a physical mobile device backup, they performed a client-side analysis on the Amazon app and AVS that revealed the last given voice commands. *Krueger and McKeown* [41] populate an Echo device with generated data and decrypt and extract the data stored on the Amazon server through the Alexa APIs and a proxy (Burp Suite) by scraping the communication from the server to the proxy. The impact of data deletion from users as a counter-forensic strategy is also examined, the conclusion being that metadata such as timestamps, location data, and device information stored by a great diversity of APIs will often lead to proof for actions, even when users try to delete them by means provided by the Alexa application. *Youn et al.* [64] examine the Echo Show, the first smart speaker with a display studied in the literature. Data is obtained from the hardware, client, and cloud level, and for the received data, the three information categories: account (e.g., username, nickname), system (system log files), and activity (e.g., surfing the internet reading e-mails) are identified. They present a forensic framework for smart display hubs as a process diagram and demonstrate its potential use in investigating a constructed criminal case. *Giese and Noubir* [23] provide an in-depth analysis of data extraction from NAND memory. In a case study for data extraction from bought used devices, they show how to obtain data from devices of different levels of reset and functionality, gaining Owner (name, username, e-mail), Wi-Fi Network (SSID, SSIDs in proximity), and MAC addresses and names of paired devices. The exploited vulnerability is also confirmed for other popular Smart Home devices. *Villarreal et al.* [58] also focus on data extraction from memory chips, relying on a non-destructive way of obtaining memory data through ISP contacts, as known methods require destructive chip-off techniques or complicated micro-soldering operations only applicable by experts. The goal was to develop a process to create easy-to-use 3D printable standard test probe jigs (device-specific fixtures to extract data with pogo pins as contacts, having identified the ISP contacts by CT scan) for any given memory card. Analyzing memory from a used Echo Dot, they obtain data similar to *Giese and Noubir* [23] and data on installed software versions, SQLite 3 databases, and logs with timestamps. *Lorenz et al.* [45] similarly focus on extraction using ISP contacts of various versions of Amazon Echo Show 1, 2, 5, 8, 10, and the Amazon Echo Spot. Further, they provide a methodology showing how to locate and identify device owner account information, device specifications and configurations, and the location of user activity on the device that may contain bare evidence and contraband. With a four-step process, they conduct experiments on IoT devices that can help law enforcement professionals prepare affidavits for search warrants with information that helps meet the legal requirement that evidence of a crime is likely to be contained on the IoT device.

The results of eight experiments conducted on Amazon IoT devices are intended to provide law enforcement with concrete, practical instructions on extracting data from Amazon IoT devices. The data extraction is extended with a background analysis of IoT device forensics that discusses the rationale for seizing digital evidence and the types of PDE, leading to a complete device diagram example that helps law enforcement understand how evidence is stored on IoT hardware through everyday interactions with the device.

Table 3. Evolution of device forensics on smart speakers in smart homes.

Source	Year	Extracted Data			Extraction Approach			Tool(s)	Device(s)
		D	U	C	Manual	Logical	Physical		
[52]	2018	○	●	○	●	●	○	N/A	Amazon Echo Dot 2
[63]	2019	○	●	●	●	○	○	N/A	Amazon Echo Plus 2, Google Home Mini
[43]	2019	●	○	●	○	●	●	CLI, Zenmap	Simulated Alexa Pi on Raspberry Pi
[35]	2019	○	○	●	●	●	●	Clova DF web proxy	AI-Speaker: Clova, NUGU Kakao Mini, GiGa Genie
[53]	2019	○	●	○	○	○	●	FTK, VNR, XWays, UFED	Amazon Echo Dot 3
[41]	2020	○	●	●	○	●	○	Burp Suite, Portswigger	Amazon Echo Dot 2 & 3
[44]	2021	○	○	●	○	●	○	Tensorflow, NLP	Xiaomi AI-Speaker
[64]	2021	●	●	○	○	○	●	EasyJtag	Amazon Echo Show 2
[23]	2021	○	●	●	○	○	●	N/A	Amazon Echo Dot 3 (factory reset)
[58]	2023	●	○	●	○	○	●	EasyJtag, RIFFBox, FCC ID Lookup	Amazon Echo Dot 2
[45]	2023	●	●	●	●	○	●	ISP	Amazon Echo Show 1,2, 5, 8, 10, Amazon Echo Spot

D = Device, U = User, C = Communication

AI Speaker. In addition to the classic smart speakers, AI speakers are becoming increasingly popular. Therefore, methodologies for forensic evidence extraction must, of course, be developed or adapted. *Jo et al.* [35] work with AI speakers; since they are usually always in use, they can provide substantial evidence for DF. For example, AI speakers have supplied proof of murders in the US and Mexico and are released without specific regulatory guidelines. This study proposes five DF analysis procedures for four Korean AI speaker models (Clova, NUGU Kakao Mini, GiGa Genie) from different manufacturers. The five methods are applied to all AI speaker models, and the results are presented. In particular, they developed a forensic tool to trace the user command history for NAVER Clova. *Lin et al.* [44] choose an unusual approach as they intend their forensic framework specifically to be used to surveil the behavior of smart speakers to prevent user privacy breaches. They use network extraction analysis by employing a man-in-the-middle attack through ARP spoofing to determine if the speaker's network traffic is similar to the traffic patterns observed in previous commands.

> **RQ3: How advanced is device forensics on smart speakers and voice assistants in smart homes?**
>
> The smart speakers and voice assistants category is dominated by research on products from the Amazon Echo line and Amazon's proprietary voice assistant Alexa. Research has developed from a rather superficial examination to data extraction methods in this area, becoming more hardware-oriented and sophisticated. The problem is transferring forensic techniques to new versions of smart/AI speakers (incl. voice assistants).

4.2 Advances on Wearables

The second big group of devices identified in research are wearables, which comprise smartwatches, glasses, and fitness trackers. The evaluation features more diverse instruments than the smart speakers and voice assistant category. The evolution of device forensics on wearables is visualized in Table 4.

Samsung Smart Watches and Fitness Trackers. *Baggili et al.* [7] are among the first to forensically examine a Samsung Galaxy S4 Active phone populated with the Samsung Gear 2 Neo watch data and the LG G watch. In addition, they present a method for physically extracting data from the watches after gaining root access to them. They can recover a large amount of PDE directly from the watches compared to the data on the phone that was synced with them. Thus, the data at the core of the functionality, namely messages, health and fitness data, emails, contacts, events, and notifications, can be retrieved directly from the captured images of the watches. *Odom et al.* [49] focus on the SamsungTM Gear S3 Frontier and the SamsungTM Galaxy S8. They show a logical and physical extraction approach on a standalone and connected (companion device) smartwatch. Although the collection capabilities were limited, it stored all identified connected data and all standalone data appropriate for mobile communications. *Becirovic and Mrdovic* [8] also emphasize their research as a physical analysis of the Samsung Gear S3. A sequence of events is executed with the watch to confirm, besides acquiring other data on the device, to what extent they can be reconstructed by forensic analysis. Files were extracted through Wi-Fi by the software installed on the watch, which was assumed to be unlocked. The work also examines the data available if the watch is synced with a phone and the limits of data extraction from a non-rooted device. The latest research on Samsung smartwatches do *Kim et al.* [39] by developing a forensic model for wearable devices, identifying user-related artifacts by applying the model to actual wearable devices (Samsung Galaxy Watch 1 and 3, Galaxy Active 2, Apple, Garmin)

Table 4. Evolution of device forensics on wearables in smart homes.

Source	Year	Extracted Data			Extraction Approach			Tool(s)	Device(s)
		D	U	C	Manual	Logical	Physical		
[7]	2015	○	●	●	○	○	●	UFED 4PC & PA, XRY, Autopsy, ADB, SDB, toybox, netcat	Samsung Gear 2 Neo, LG G watch
[54]	2017	○	●	●	○	●	●	Linux CMD, ADB, Sleuth Kit SGGFT, MTKII	Google Glass
[2]	2018	○	●	●	●	●	○	UFED 4PC	Apple Watch Series 2
[1]	2018	○	○	○	○	●	●	UFED PA, iExplorer	Apple Watch Series 1
[47]	2019	○	○	●	○	●	○	FTK Imager, Autopsy, GoldenCheetah, FitSDK	Garmin forerunner 110, Fitbit Charge HR, HETP fitness tracker
[24]	2019	●	●	●	●	○	●	FundoWear, BTNotification, FlashTool, UFED PA, R-Studio	Fitness trackers: No. 1 G6, Cawono DZ09, Kingwear GT08
[49]	2019	○	●	●	○	●	●	UFED 4PC & PA, GrayShift's, GrayKey, SDK, SDB, X-Ways, Xcode	Samsung Gear S3 Frontier, Apple Watch Series 3
[8]	2019	○	●	●	○	○	●	SDK, SDB	Samsung Gear S3 Frontier
[36]	2020	○	●	○	○	●	○	MD-NEXT, Python, Google Earth	Xiaomi Mi Band 2, Fitbit Alta HR
[1]	2020	●	●	●	○	●	○	VMWare, Sleuth Kit, Autopsy, QEMU, Bulk Extractor	Fitbit Alta, Fitbit Ionic
[17]	2021	○	●	●	○	●	●	ttwatch, linux CMD, UFED 4PC & PC, FTK Imager, DB Browser, Runalyze	Tom Tom Spark 3
[59]	2021	●	●	○	○	●	●	UFED, MSAB XRY, Genymotion	Fitbit Versa
[29]	2022	○	●	●	○	○	●	Magnet AXIOM, UFED, Whisper App	Amazon Halo Band, Garmin Vivosmart 4, Mobvoi TicWatch S2
[39]	2023	○	●	●	●	●	●	Odin, ADB, SDB, checkra1n, UNetbootin, PuTTY, iBUS, Xcode	Galaxy Watch 1, 3, Active 2, Apple Watch Series 3, 5, Garmin Vivosport
[19]	2023	●	●	●	●	●	○	ADB, DBBrowser-forSQLite, litecli, HxD, Python.JSON, ent, Autopsy, CyberChef, VS	Xiaomi Mi Band 6
[34]	2023	●	●	●	●	●	○	ADB, MD-NEXT, FileZilla	Galaxy Watch 4, 5, 5 Pro

D = Device, U = User, C = Communication

to confirm its applicability. Meaningful data such as call and text messages, voice assistant records, media files, reminders, and health data were collected using Samsung smartwatches with user privileges. However, because health data was encrypted, it could not be analyzed. *Jeon et al.* [34] discuss the investigation of the Samsung smartwatches Galaxy watch 4, 5, and 5 Pro. To extract data on the watches, users, and communication, the authors use ADB, dumpsys, and MD-NEXT to examine application operation records and network connections thoroughly. However, the analysis becomes ambiguous over time due to the short lifespan of logs. They notice that while Android smartphones can be connected via USB for a precise analysis of companion apps, the reviewed smartwatches lack a USB port, limiting the scope of investigations.

Apple Smartwatches and Fitness Trackers. *Alabdulsalam et al.* [2] describe as one of the first forensic approaches for an Apple watch series 2. To do so, they apply a logical and manual acquisition approach. To get the file system (logical acquisition), they use the software UFED 4PC. After that, a manual investigation is performed to determine what data is stored on the watch. This method of investigation was chosen because no physical access was possible and to prove that the watch not only generates data but is also stored directly on the watch. Finally, they analyze forensic artifacts (e.g., pictures, e-mails, calendars, contacts, GPS data, health app data) retrieved from the Apple watch. In the same year *Al-Sharrah et al.* [1] provide technical details and architecture of the Apple Watch Series 1 and propose a forensic framework for smartwatches, consisting of a physical analysis of the watch combined with an initial backup to spot PDE modification, do a backup analysis of the connected phone, and wireless communication analysis of the watches' Bluetooth and Wi-Fi connections. The framework is then demonstrated on an Apple Watch Series 1, while wireless communication analysis is omitted. The backup of a paired iPhone is then analyzed in a second step, revealing sensitive information about the watch like serial number, Wi-Fi, and Bluetooth MAC address. *Odom et al.* [49] show a logical and physical extraction approach on standalone and connected (companion device) smartwatches. The Apple Watch Series 3 analysis shows that more needs to be done to determine its probative value for DF. Current collection methods are limited to manual extraction, which not only does not provide additional data on the accompanying cell phone but also provides time-sensitive or incomplete data. The latest research on Apple watches do *Kim et al.* [39] by developing a forensic model for wearable devices. Since the applicable forensic methods differ depending on the device, the forensic model was divided into logical and physical extraction based on the wearable device ecosystem. The proposed model was applied to wearable devices (Apple, Samsung, Garmin) to confirm its applicability. An Apple Watch Series 3 and 5 (GPS/GPS+Cellular) are used for the analysis. All devices are accessible by Wi-Fi, enabling a PC connection method. As a result, the forensic model produced the same results, and the same artifacts were derived for the tested models by utilizing a manual, logical, and physical extraction approach.

Garmin Smartwatches and Fitness Trackers. *MacDermott et al.* [47] conduct as one of the first forensic analyses on a Garmin Forerunner 110. The authors identified forensically relevant artifacts by connecting the wearable directly to a Windows 10 computer. They noted that the Garmin device recorded a wealth of PDE. Specifically, they could recover details about their test runs to populate the devices with data, deleted records, and user and device information. However, the authors did not investigate what PDE could be recovered from the mobile apps of these devices, which we believe is a limitation. *Hutchinson et al.* [29] analyze wearable-related applications for Android mobile apps (incl. Garmin Connect) being connected to a fitness tracker of the same brand (Garmin Vivosmart 4). After populating the Vivosmart by wearing, the information stored in the synchronized apps was extracted by creating a backup of the rooted Android phone. The backup contained pertinent data for criminal investigations regarding the user's location, health, and activity for all three devices. Unsuspected recording of location-related data like names of nearby weather stations, IP addresses, or city/county names was named as a privacy concern yet a valuable source for investigations. The latest research on Garmin watches or fitness trackers do *Kim et al.* [39] by developing a forensic model for wearable devices. Since the applicable forensic methods differ depending on the device, the forensic model was divided into logical and physical extraction based on the wearable device ecosystem. The proposed model was applied to wearable devices (Apple, Samsung, Garmin) to confirm its applicability. For the analysis, a Garmin Vivosport Smartband has fewer features than a smartwatch. Thus, the Vivosport communicates with the cloud server by pairing it with a smartphone via Bluetooth. Various artifacts could be recovered using a manual, logical, and physical extraction approach (user, health, and location information).

Fitbit Smartwatches and Fitness Trackers. *MacDermott et al.* [47] conduct as one of the first forensic analyses on a Fitbit Charge HR. The authors identified forensically relevant artifacts stored by these devices by manually examining the associated Windows 10 app. They noted that the Fitbit Windows 10 app recorded a wealth of PDE. Specifically, they were able to recover details about the test runs they conducted to populate the device with data, including GPS locations, group chat/post interactions, and other user and device information. However, these authors did not investigate what forensically relevant data can be recovered from the mobile apps of these devices, which we believe is a limitation. *Kang et al.* [36] perform forensic analysis on the Fitbit apps with an Android 7 device after using the Fitbit Alta HR fitness tracker. The authors show that it is possible to recover artifacts related to user-entered profile information (e.g., date of birth, name, weight, and height), the device used (device MAC address and ID), daily sleep and step records, and activity data (step count and distance timestamp). Concerning the Fitbit app, they were able to recover device information, step counts, sleep information, activity information, and GPS data associated with a tracked exercise. The limitation of this work was the need for more information on how the fitness trackers were populated with data. *Almogbil* *et al.* [4] did a forensic analysis of Fitbit's Windows 10 desktop app after populating the app with data from an Ionic smartwatch and an Alta fitness tracker. The authors focused on guiding forensic investigators in navigating Fitbit data stored in the desktop app, distinguishing between manually entered and automatically recorded data and performing a quick but thorough analysis of Fitbit health data using open-source tools. *Williams et al.* [59] perform a forensic analysis of the Fitbit Versa smartwatch running Android 9 and iOS 12. After applying logical and physical extraction methods, the authors clearly described the relevant user artifacts recovered from these two devices. The authors also comprehensively compared the forensic acquisition capabilities of UFED and MSAB XRY.

Low-Cost Smartwatches and Fitness Trackers. *MacDermott et al.* [47] conduct as one of the first forensic analyses on a generic low-cost HETP fitness tracker. The authors tried identifying forensically relevant artifacts by manually examining the mobile app. However, the authors could not view data that may have been stored in the app and thus were unsuccessful in retrieving PDE. *Gregorio et al.* [24] highlight that low-cost smartwatches move away from high-end smartwatches with standard OS and state that current forensic tools have minimal support for these types of devices, leaving the low-cost smartwatch devices behind, even though they could provide valuable PDE. Thus, the authors apply a manual and physical acquisition on several no-name fitness trackers: No. 1 G6, Cawono DZ09, and Kingwear GT08, all equipped with a low-cost MTK chip. For data acquisition, a non-forensic tool (FlashTool) is applied. To quickly search for information in the entire unknown structured data of Nucleus RTOS-based smartwatches, various commercial forensic tools were utilized to validate the digital traces. *Hutchinson et al.* [29] investigate the Mobvoi App, connected to a Mobvoi TicWatch S2. After populating the armband by wearing and using it, the information stored in the synchronized apps was extracted by a backup of the rooted Android phone. The backup contained pertinent data regarding the user's location, health, and activity (primarily health-related PII Data for the Mobvoi device).

Xiaomi Smartwatches and Fitness Trackers. *Kang et al.* [36] examine the Xiaomi Mi Band 2, designed to monitor body, sleep, and activity. The records generated by these features can be crucial for criminal investigations. Fitness wristbands interact with mobile companion devices. This allows relevant data to be collected from the connected device. The trackers manage most of the data in SQLite3 databases for Android devices. A logical extraction is done to get there by extracting an image of the user data partition from the companion device with the Mi Fit App (mobile forensics software MD-NEXT). This way, user and communication data could be revealed. *Domingues et al.* [19] enhance the previous approach and analyze the forensic artifacts stored by the ZeppLife (former Mi Fit) application on an Android mobile device paired with a Xiaomi Band 6. The Mi Band 6 is populated with heart rate, SpO2, sleep duration, pedometer, and exercise data. Some data is stored in the cloud and can be accessed with credentials. The ZeppLife application stores much of the data in XML files and SQLite3 databases. Valuable artifacts include activity/rest/rescue times, steps, oxygen saturation values, alarms, reminders, GPS coordinates, and timestamps in UTC. They developed various python apps to facilitate the use of MiBand6/ZeppLife for forensic analysis.

TomTom Smartwatches and Fitness Trackers. *Dawson and Akinbi* [17] analyze the TomTom Spark 3 GPS fitness watch. They investigate the storage locations and identify and extract forensically interesting artifacts on the physical smartwatch using Ttwatch. In addition, they identify and reconstruct PDE related to user information, past activities, and GPS locations generated by the smartwatch and stored in databases by the TomTom Sports mobile app installed on an Android smartphone (with UFED). They identified proprietary activity files (.ttbin) containing PDE related to user activities stored on the Android file system and the physical smartwatch. Using the Runalyze web platform's non-forensic tool for athlete performance analysis, revealing activity files and past user activities (incl. GPS locations).

Google Glasses. *Rongen and Geradts* [54] investigate Google Glass with different extraction methods (software- and hardware-based). Afterward, the extracted data is examined for possible artifacts left behind during regular use. They show that a lot of forensically interesting information can be retrieved from Google Glass (e.g., images, videos, contacts, social media activity, locations and destinations (navigation), audio (including user voices and TTS cache), interactions from various messaging platforms, sync information (e.g., for Google+), connected devices, Android Wear interactions, web browsing behavior, search history, and calls).

> **RQ3: How advanced is device forensics on wearables in smart homes?**
>
> Almost all wearable devices have data extracted through some backup from a companion device/app, except for [8], which extract data directly via a Wi-Fi transmitted backup. Traffic analysis or decryption is not conducted, and data is not pulled directly from chips, showing a much more limited repertoire of techniques than smart speakers. However, this is due to the much smaller design of wearables, which makes access to chips more difficult. In addition, the market share by vendor from 2022 [42] shows the following breakdown: Apple with 30%, and Others (no-name brands) with 29%, then Samsung with 10%, Huawei with 7%, Amazfit and Garmin with about 4%. This distribution differs from the research conducted in wearables, which still leaves some gaps, especially in the no-name devices, Huawei and Amazfit.

4.3 Advances on Smart Things

The third category that could be identified are smart things, including all devices that do not fit into the other categories. Thus, devices in this category range from sensors, hubs, lamps, cameras, plugs, and locks to intelligent scales, TVs, and robot vacuum cleaners. The evolution of device forensics on smart things is viusalized in Table 5.

Smart TV or TV Stick. *Boztas et al.* [10] propose new procedures for acquiring, analyzing, and investigating a smart TV. The flash memory on the investigated Smart TV is an eMMC chip. Therefore, a logical (software method, file system) and physical extraction (eMMC five-wire method, NFI Memory Toolkit II) approach have been applied. Since data acquisition by chip extraction is very specialized and destructive, a software method was investigated. The software method is vulnerable to updates of the firmware of the Smart TV. Therefore, the authors highlight that hardware methods for accessing Smart TVs will be more lasting and forensically sound than software. *Hadgkiss et al.* [26] investigate the first and second Amazon Fire TV stick generation. The second generation supports USB devices and storage media through the HDMI port, while the first generation requires it to be rooted before data can be extracted. In addition, extraction via ADB, the UFED Touch Test, rooting, custom Python scripts, and manual extraction are applied. Some methods provided data, but forensic reliability was compromised in most cases. The authors, therefore, applied a physical acquisition approach (chip-off), leading to relevant evidence data (e.g., user info, device name, Wi-Fi info, Amazon video content).

Smart Cameras. *Bhardwaj et al.* [9] forensically analyze a smart IoT camera device commonly found in households, the D-Link DCS-5020L Wireless Network Camera smart IoT surveillance camera. This IoT device works day and night and offers several features, such as a wide field of view, 4x digital zoom, motion detection, 120-degree pan and tilt, a mobile app, and remote access. The device also has a built-in wireless extender that can extend Wi-Fi coverage up to 8 m. During the examination of the logically extracted firmware from the camera, the authors found some folders with details of "admin", "root" and "password", "passwd". These files pose a high risk as they contain hard-coded usernames, passwords, and email addresses. The authors even found IP addresses and URLs. *Kim et al.* [40] review the KASA smart home camera forensically, as it has a built-in microphone and speaker enabling two-way communication through the camera and its companion app. Thus, they extract movement from the house and connect it to a specific individual. For this, they solely apply manual extraction of PDE with the tools CURL and an online decoder.

Smart Lights and Switches. *Do et al.* [18] provide a forensic analysis on the LIFX Original 1000 (smart light) connected to a local Wi-Fi network to receive commands from any compatible device. The second device is Belkin's WeMo (smart switch), which allows a user to control the power state of a connected device (via App). Since a user must explicitly control the app, capturing communication between the phone and the Belkin WeMo is considered infeasible for

Table 5. Evolution of device forensics on smart things in smart homes.

Source	Year	Extracted Data			Extraction Approach			Tool(s)	Device(s)
		D	U	C	Manual	Logical	Physical		
[10]	2015	●	○	●	○	●	●	MTK II	Samsung smart TV
[20]	2018	○	●	●	○	○	●	FEAAS, DB Browser, PLIST Editor, iTunes, SQLite3, Epoch Converter, ChromagnonCache	Nest thermostat, indoor/outdoor camera, Google Home Mini
[6]	2018	●	○	●	●	●	●	UFED Touch 2, SSH, Filza, iFile	Almond+ Hub with: Philips Hue lamps, Jasco/Securifi plugs, Fibaro/NYCE
[18]	2018	●	○	●	○	●	○	Broadcast UDP, (ARP scanner)	LIFX 1000 light, Belkin WeMo switch
[26]	2019	●	●	●	○	○	●	FTK Imager, WinHex, X-Ways, DB Browser, XML Marker, VMware, Linux Mint	Amazon Fire TV stick
[11]	2019	○	●	●	○	●	●	FTK Imager, KAPE, Autopsy, QPhotorec	Windows 10 IoT Core
[40]	2020	○	●	●	●	○	○	Curl, online decoder	Google nest hub, TP-Link Kasa camera, Samsung smart things
[31]	2020	○	○	●	○	●	●	Ettercap, ADB, kpartx, SQLite, wireshark	Various Smart Plugs
[30]	2020	○	●	●	○	○	●	Magnet ACQUIRE & AXIOM, MSAB XRY	August smart lock, doorbell pro
[61]	2020	○	●	●	○	○	●	ToolKit (FTK), OS Forensics	Smart Hello Barbie
[55]	2020	●	○	●	○	●	●	JTAG, ISP, TTL, UART, EtterCap, tcpdump, Avrdude, GHex, Debian, SleuthKit, Autopsy, Radare2, esptool	Raspberry Pi 3 Model B, Arduino Uno rev. 3
[25]	2021	○	●	●	●	●	●	Ubuntu Linux, Flashrom, MSB XRY	iHealth body scale
[65]	2022	○	●	●	○	●	●	Magisk, TWRP, ADB, Kali Linux	Roomba j7+
[12]	2022	○	○	●	○	○	●	N/A	Xiaomi Mi sensor kit
[38]	2023	●	○	●	○	●	○	CARPE, Python, Proof-of-Concept tool	Chromecast 3rd, Google Home Mini, Induction Range, Sorento 2016 Gen., Yeelight Smart LED Bulb 1S
[22]	2023	○	○	●	○	●	○	AE forensic (PoC tool)	Raspberry Pi, Enviro pHAT light & temperature sensor
[9]	2023	●	●	●	○	●	○	Binvis, Binwalk, Firmwalker, VirusTotal API, Entropy	D-Link DCS-5020L wireless network camera

D = Device, U = User, C = Communication

an attacker. This contrasts with the LIFX Original 1000, where the smartphone service repeatedly communicated with the device, producing more data (e.g., Wi-Fi and device information). The device's state may be the most important information a forensic investigator can obtain from a smart switch.

Smart Plugs. *Iqbal et al.* [31] explore various smart plugs (Mydlink, Kasa, Telldus, Live, and Alexa) to conduct forensic analysis. They extracted a smartphone copy (logical acquisition) using ADB. Only limited evidence of the smart plug activities could be found on the smartphone. Compared to other devices, they concluded those do not provide much forensic evidence (e.g., activate, deactivate, configure, register an account).

Smart Lock and Doorbell. *Hutchinson et al.* [30] set up their own mini-smart home, which includes a TP-Link router, an Alfa network adapter that allows capturing all network traffic, three smartphones (one iPhone 7, two Samsung Galaxy S7), and an August Smart Lock with Doorbell. The authors chose to forensically capture each smartphone device at three points: a baseline image, an image after interacting with the devices, and an image after rooting/jailbreaking the device. This enabled the identification of several artifacts, like owner information, guest devices, and device interactions, and downloaded and cached images from the doorbell, which are all recoverable on iOS and Android devices, albeit to varying degrees. Accurate user GPS data can only be recovered from a rooted Android device.

Smart Body Scale. *Grispos et al.* [25] analyze an IoT ecosystem of an iHealth wireless body scale, including two phones with installed iHealth apps and a desktop computer used to access the iHealth cloud portal. The scale was used for several days to populate it, the cloud portal, and the other devices in its environment with data. Data were extracted from the scale memory chips physically with a breakout board, from the mobile devices using commercial forensic software to create data images and visually from the cloud portal. Weight measurements, timestamps, and various body measurements could be extracted from all three sources.

Smart Robot Vacuum Cleaner. *Zhou et al.* [65] examine a robot vacuum system. The associated devices examined are the Roomba j7+ and a rooted Android mobile device with the iRobot Home App. Tools and the process for rooting the phone and extracting an image file from it are thoroughly explained, while the process for populating the device with data is omitted. The image file is then searched for data from the iRobot App, revealing, e.g., app usage details, including timestamps, activity, cleaning schedule, user credentials, and network information.

Sensor Kits. *Castelo Gómez et al.* [12] conduct a fornesic analysis on the Xiaomi sensor set and the environment it builds. Since the Xiaomi sensors (window, door, motion, and wireless switch) only establish communication via the Zigbee protocol, remote access can be ruled out. At the same time, these sensors only have memory in the Zigbee module soldered to the circuit board, with only 512 KB of integrated flash memory. Therefore, the only available data extraction methods are JTAG/UART or chip-off. Though these methods may be theoretically compatible, they might not be the optimal approach to pursue, as the authors failed to successfully implement them. Moreover, choosing either remote or live acquisition is unattainable when the devices have the most recent firmware installed. The authors highlight that firmware modifications can greatly impact IoT forensic investigations. They found that all data generated by the environment is sent to the Xiaomi cloud, enabling access to it with the smartphone app. Moreover, it was established that certain types of data are stored in the memory of the hub or the sensors, provided these can be accessed.

Smart Connected Toys (SCT). *Yankson et al.* [61] examine current DF solutions for SCTs and present a forensic investigation framework for investigators to successfully use Plan, Preserve, Process, and Present (4P) to conduct a DF analysis of an SCT in a situation where the SCT is involved or is the subject of a crime. The authors were able to perform SCT forensics on the Hello Barbie. However, no easy-to-use forensic tools currently allow data extraction from the Hello Barbie device. For example, the Wi-Fi scan results showed an open SSID. The availability of such an SSID may pose a security risk (spying).

Complete IoT System. *Dorai et al.* [20] conduct a forensic analysis of an intelligent home IoT ecosystem. They use the third-generation learning Nest thermostat, a Nest indoor and outdoor camera, and a Google Home Mini speaker to do this. The researchers developed an open-source forensics tool called Forensic Evidence Acquisition and Analysis System (FEAAS) that aggregates evidence data into a readable report that can be used to make inferences about user events. It was possible to find relevant data from databases recovered from the mobile device backup (date and time when specific changes were made). This includes whether someone manually calibrated the thermostat, whether the user was home at a particular time, whether the user spoke to the Google Home Mini device, and whether the camera was intentionally turned off at a particular time. *Awasthi et al.* [6] do a forensic analysis on the Almond+ ecosystem (including the Home Hub, iOS/Android companion apps, and cloud environment) to provide a method for extraction and analysis. The Almond+ ecosystem has several local and cloud-based protocols available for analysis. Information is also available through a companion app that provides evidence of connected network devices and user activity. Observed changes such as humidity, temperature, and motion are more passive in their approach. Data collected by an Almond+ system could provide a comprehensive, albeit text-based, picture of events leading up to a crime scene. *Castelo Gómez et al.* [11] conduct a forensic analysis of the

Windows 10 IoT Core OS non-volatile memory, providing guidelines for implementation and detailing aspects such as the analysis, collection, and evaluation of the evidence found. This provides investigators with a study that they can use as a tool when investigating the same OS. In addition, a module for KAPE, a forensic program, has been developed to collect all relevant sources of information stored in Windows 10 IoT Core's non-volatile memory (e.g., timeline creation, user files, antivirus logs). This allows investigators to automate the search for evidence in future investigations. *Kim et al.* [40] examine the data that can be collected from smartphones, the Google web interface, and private Google Home APIs. Companion apps installed on smartphones and paired with smart home devices can provide information about the location and model of each device. The activity history recorded by each device may be helpful for forensic investigations. For example, a suspect's statement can be verified using movement, temperature, command, and voice data, which can also be used to infer the time of an event. This study identified the sources of different data types generated by a certain number of smart home devices, which can be compared to determine the actual sequence of events. *Shalaginov et al.* [55] provide a step-by-step evidence analysis in a smart home environment. The setup contains a wireless router and an IoT device with sensors (Raspberry Pi 3 model B). They investigated the IoT device with live on-chip access via JTAG, ISP, and TTL. Additionally, they use a forensic workstation equipped with EtterCap to apply ARP poisoning to capture and record (using tcpdump) the traffic between the router and the device. Using esptool, they were able to read the flash memory over the UART interface. In addition, they could acquire a read-only image from the IoT Hub. Thus, they could extract timestamps, network traffic, and firmware images. *Kim et al.* [38] provide a study on connected IoT services by examining forensic artifacts in six scenarios. The authors found that not all services show traces of connectivity. Based on the scenario findings and existing forensic IoT frameworks, they propose an improved forensic IoT model to identify the connectivity between things, ensure relevant data collection, narrow the areas of investigation, and integrate the data. They focus on correlating contextual information and visualizing it to support forensic investigators. The tool interprets the data from a target source, the names of connected sources, which may represent services, devices, or applications. All identified destinations and connected sources are represented by nodes, which contain an identifier and label, and edges connecting these nodes. *Gandhi and Arumugam* [22] provide an attempt to standardize the process of extracting electronically stored information from connected embedded devices by designing AE forensics and supporting it in IoT device architecture. The authors test their tool on a conceptual implementation, using oneM2M deployment, &CUBE IoT device software, nCube-Rosemary as an open-source IoT gateway platform, and as devices a Raspberry Pi with nCube-Thyme software, Enviro pHAT light and temperature sensors. Thus, they can forensically prepare a system with their software AE forensics and standardize the extraction of electronically stored information from a connected embedded device by-design.

> **How advanced is device forensics on smart things in smart homes?**
>
> The approaches here are more diverse again, including fundamental network traffic analysis, logical extraction of a companion device, like a smartphone or direct extraction from memory chips (chip-off). The category of Smart Things shows that the more negligible the IoT devices and the less memory they have of their own, the more dependent they are on a companion device (e.g., smartphone, hub) and the associated app that provides a smartphone or hub image plays a vital role in almost all papers.

5 Discussion and Open Research

The field of smart home IoT device forensics is fraught with unique challenges that stem from the inherent characteristics of IoT devices and the environments in which they operate. Some of these are similar to the ones found in IoTF generally, while others are specific to device forensics. Against our expectations, we could rarely identify challenges that are specific to only one category. However, these challenges also present opportunities for future research.

5.1 Specific Challenges

We could identify a few challenges specific to a device category during the systematization of current research. Starting with (1) Smart Speakers and Voice Assistants the challenges of interpreting audio [41], the analysis of the users intent as well as the sound assignment of a recorded voice to a person (voice falsification) [44] are discussed. In the category (2) Wearables, authors discuss challenging aspects like the interpretation of GPS and health data [2,19,39]. In addition, the current lack of research on low-cost IoT products leads to unsuitable methods, making it hard to recover PDE [17,24]. The third category (3) Smart Things also focuses on interpretation problems regarding energy readings [33], and video material [26,54]. Interpreting digital traces is not specific to IoT devices but is a very actual problem. As lately discussed by *Horsman* [28], who emphasizes the importance of accurate interpretation of digital traces in DF. Given the growing reliance on PDE, he proposes an 'eight-pillared' framework to ensure quality in interpretive work.

5.2 Challenges per Investigation Phase

Whereas there are only a few challenges specific to the device categories, some common patterns represent challenges we could observe over-spanning the categories, Smart Speakers and Voice Assistants, Wearables, or Smart Things. Instead, these overarching challenges that represent problems existing in IoT device forensics in smart homes can be discussed based on the in Sect. 3 defined investigative phases in IoT environments: (1) Initialization, (2) Acquisition, and (3) Investigation.

Initialization Phase. Comprehending the complete IoT ecosystem is crucial in identifying PDE sources related to an incident [43]. The data derived from IoT devices hold significant value and relevance in forensic investigations, which seems to be not always known [1,7,17,24,46]. Furthermore, identifying the devices present during the incident is essential in IoT device forensics [43,45].

Acquisition Phase. The use of forensically verified tools is essential for soundly collecting IoT devices and data [2,17,30,45,47]. Non-invasive data acquisition methods are employed to retrieve data without altering or damaging the evidence. The process of locating IoT devices for data collection and extraction can be challenging due to their physical location, but apps on companion devices might help [7,11,45]. Accessibility issues go in the same direction and arise mainly when accessing IoT devices and their data for forensic analysis because the data is not stored directly on the device but uploaded to the cloud. This issue is further complicated by the involvement of multiple jurisdictions [6,11,17,19,20,35,63]. Moreover, there is an urgency in capturing and preserving IoT data promptly before it becomes unavailable [8,35,43]. To obtain the maximum amount of data, a challenge can be that the username and password are primarily necessary to apply the preferred investigation method [19,59].

Investigation Phase. Live forensics involves the real-time analysis of collected data to determine incident-related details, which can be challenging but crucial in device forensics [18,35,43]. This can support timeline extraction, which reconstructs chronological sequences from disparate IoT data during analysis [8,17,41]. Moreover, the integrity of IoT data is to be ensured throughout the investigative process, maintaining its accuracy and trustworthiness, which can be difficult during IoTF [35]. Additionally, evidence admissibility is crucial, ensuring that the collected IoT data meets legal standards for the investigative report. The complexity of IoT data structures and diverse technologies involved in the analysis are hard to handle effectively [6,55]. Machine learning models can be incorporated for practical analysis of IoT-derived data, but also endanger traceability [55]. Often, a combination of various tools and approaches is vital in viewing and interpreting the acquired evidence and deciding how to proceed, but which tools to use can be challenging for IoT investigators [47].

5.3 Overarching Challenges

Here identified challenges are discussed that can be found across all investigation phases. Starting with anti-forensics, which involves tackling challenges related to attempts that manipulate, obfuscate, or hinder IoTF investigations [41,43,63]. Another crucial aspect in device forensics is managing encrypted data and its impact on forensic analysis [17,23]. Challenges arise from the diversity in IoT hardware and software ecosystems, including different protocols and data formats, which complicates a standardized approach to extract and analyze PDE [6,17,18,36,47]. Issues associated with smaller-sized IoT devices and their impact on forensic procedures are also considered [39]. The diversity of IoT devices makes it challenging and shows the need for standardized protocols and formats within the diverse IoT ecosystem [18,30,45,55]. Further, concerns regarding data privacy and protecting sensitive information are addressed [8,35]. Overall stages of an investigation, the availability, limitations, and suitability of tools specifically designed for IoT device forensic analysis are crucial challenges [2,17,18,43]. Issues related to intentional or unintentional tampering or manipulation of IoT data are addressed [7,8,10,17,52]. Legal issues generally span over all layers in IoTF, including the devices, and the differences in legislation (per country) challenge investigations [19,20,35]. These devices' forensic collection and analysis has become a significant challenge for investigators due to how much data they generate [6,17,30]. A combination of multiple challenges discussed before makes it hard to develop a generally applicable approach or model for digital investigations in IoT environments [2,17,61]. Ancillary, there needs to be more education and training for forensic investigators, experts, and learners in IoTF generally, as well as all subdomains, like device forensics [30].

5.4 Open Research

Generally, the smaller the IoT device gets, the more dependent the forensic analysis is on the companion devices and apps to extract PDE. While research on smart speakers progressed continuously from superficial to in-depth approaches, the transfer to new speaker versions is unclear. Another exciting insight is that the research on wearables does not replicate the actual usage pattern in society, leaving a massive gap for no-name brands. While we could present a novel view of the state-of-the-art IoT device forensics in smart home environments, we can determine that the basis for this research area is solid. However, still, several aspects require further and more in-depth research.

Collaborative Forensics is a crucial component where the forensic investigator works in tandem with various stakeholders, including the vendor, the developer, and the cloud provider. This collaborative approach is becoming integral to the investigation process, fostering a more comprehensive and effective analysis [55]. This collaboration extends into *Artifact Comparison*, where future work in device forensics should involve a meticulous comparison of the artifacts recovered using different approaches from various devices [29]. This comparative analysis can provide valuable insights into the effectiveness of different forensic methods and tools, thereby enhancing the overall quality of investigations. *Forensics Readiness* is another critical aspect, emphasizing the importance of being prepared and equipped to conduct forensic investigations efficiently and effectively. This readiness is closely tied to both collaborative forensics and artifact comparison, as it involves the active participation of all stakeholders and using the most effective forensic tools and methods [38,55]. In the context of *Crime Prevention*, the insights gained from forensic investigations (e.g., standard password, no update possible, no encryption of communication) can be used to develop more effective strategies for preventing future crimes [55]. This preventative approach is closely linked to the concept of *Forensics-by-Design*, which involves designing systems and processes with forensic considerations in mind from the outset.

6 Conclusion

As the adoption of IoT devices within in smart home environments still rises, the attack surface broadens for malicious actors. This inevitably results in an increasing need of forensic investigations of smart home devices to solve cybercrimes. While we have reviewed, classified and categorized the current state of the literature, we could extract a continuous increase in publications to date. Further, were we able to extract a taxonomy that shows what smart home device types are currently investigated, how the evidence extraction is done and what evidence data types are retrieved. Regarding the three identified main categories, we could deduct following conclusions.

- *Smart Speakers and Voice Assistants.* This field mainly focuses on studies related to Amazon's Echo products and the proprietary voice assistant Alexa. The research has evolved from surface-level analysis to more advanced data extraction techniques, shifting towards a focus on hardware and increased complexity. Adapting forensic methods to newer models of smart speakers or AI assistants poses a challenge.
- *Wearables.* On most wearable devices data is obtained through a companion device or app backup, with the exception of [8], which directly extract data via Wi-Fi backup transmission. Unlike in the category smart speakers and voice assistants, no traffic analysis or decryption takes place, and data is not extracted directly from chips, showcasing a more limited range of techniques. This limitation is attributed to the smaller design of wearables, making chip access more challenging.
- *Smart Things.* The methods presented vary once again, ranging from basic network traffic analysis to the systematic extraction of a companion device such as a smartphone or direct extraction of memory chips (chip-off). The Smart Things category indicates that IoT devices with limited memory rely heavily on a companion device (e.g., smartphone, hub) and the corresponding app, which is crucial to conduct a forensic investigation in most papers. When analyzing complete IoT systems all available devices in a smart home can be relevant in connection with associated companion devices, apps, IoT hubs and cloud environments.

Across all categories, various challenges and research opportunities can be identified. These findings highlight the significant demand for research, which may further increase with the integration and utilization of emerging technologies like large language models in IoT settings and on IoT devices. We anticipate advancements in smart home device forensics and aim to address existing research gaps concerning readiness for future investigations in IoT environments.

References

1. Al-Sharrah, M., Salman, A., Ahmad, I.: Watch your smartwatch. In: 2018 International Conference on Computing Sciences and Engineering (ICCSE), pp. 1–5 (2018)
2. Alabdulsalam, S., Schaefer, K., Kechadi, T., Le-Khac, N.-A.: Internet of Things forensics – challenges and a case study. In: DigitalForensics 2018. IAICT, vol. 532, pp. 35–48. Springer, Cham (2018). https://doi.org/10.1007/978-3-319-99277-8_3
3. Alahmadi, S., Rojas, P., Idriss, H., Bayoumi, M.: Taxonomy of consumer and industrial iot. In: SoutheastCon 2023, pp. 418–424 (2023)
4. Almogbil, A., Alghofaili, A., Deane, C., Leschke, T.: Digital forensic analysis of fitbit wearable technology: an investigator's guide. In: 2020 7th International Conference on Cyber Security and Cloud Computing (CSCloud), pp. 44–49 (2020)
5. Atlam, H., Hemdan, E., Alenezi, A., Alassafi, M., Wills, G.: Internet of things forensics: a review. Internet of Things **11**, 100220 (2020)
6. Awasthi, A., Read, H.O., Xynos, K., Sutherland, I.: Welcome PWN: almond smart home hub forensics. Digit. Investig. **26**, S38–S46 (2018)
7. Baggili, I., Oduro, J., Anthony, K., Breitinger, F., McGee, G.: Watch what you wear: preliminary forensic analysis of smart watches. In: 2015 10th International Conference on Availability, Reliability and Security, pp. 303–311 (2015)
8. Becirovic, S., Mrdovic, S.: Manual iot forensics of a samsung gear s3 frontier smartwatch. In: 2019 International Conference on Software, Telecommunications and Computer Networks (SoftCOM), pp. 1–5. IEEE (2019)
9. Bhardwaj, A., Kaushik, K., Bharany, S., Kim, S.: Forensic analysis and security assessment of IoT camera firmware for smart homes. Egypt. Inf. J. **24**(4), 100409 (2023)
10. Boztas, A., Riethoven, A., Roeloffs, M.: Smart tv forensics: digital traces on televisions. Digit. Investig. **12**, S72–S80 (2015)
11. Castelo Gómez, J.M., Roldán Gómez, J., Carrillo Mondéjar, J., Martínez Martínez, J.L.: Non-volatile memory forensic analysis in windows 10 IoT core. Entropy **21**(12), 1141 (2019)
12. Castelo Gómez, J., Carrillo-Mondéjar, J., Martínez Martínez, J., Navarro García, J.: Forensic analysis of the xiaomi mi smart sensor set. Digit. Investig. **42–43**, 301451 (2022)
13. Chung, H., Park, J., Lee, S.: Digital forensic approaches for amazon alexa ecosystem. Digit. Investig. **22**, S15–S25 (2017)
14. CNN. Arkansas judge drops murder charge in amazon echo case (2017). https://edition.cnn.com/2017/11/30/us/amazon-echo-arkansas-murder-case-dismissed/index.html. Accessed 09 Apr 2023
15. CNN. Cops use murdered woman's fitbit to charge her husband (2017). https://edition.cnn.com/2017/04/25/us/fitbit-womans-death-investigation-trnd/index.html. Accessed 09 Apr 2023
16. Davis, B., Mason, J., Anwar, M.: Vulnerability studies and security postures of IoT devices: a smart home case study. IEEE Internet of Things J. **7**(10), 10102–10110 (2020)
17. Dawson, L., Akinbi, A.: Challenges and opportunities for wearable IoT forensics: Tomtom spark 3 as a case study. Forens. Sci. Int. Rep. **3**, 100198 (2021)
18. Do, Q., Martini, B., Choo, K.K.R.: Cyber-physical systems information gathering: a smart home case study. Comput. Netw. **138**, 1–12 (2018)

19. Domingues, P., Francisco, J., Frade, M.: Post-mortem digital forensics analysis of the Zepp life android application. Forens. Sci. Int.: Digit. Investig. **45**, 301555 (2023)
20. Dorai, G., Houshmand, S., Baggili, I.: I know what you did last summer: your smart home internet of things and your iphone forensically ratting you out. In: Proceedings of the 13th International Conference on Availability, Reliability and Security (ARES 2018). Association for Computing Machinery, New York (2018)
21. Fukami, A., Stoykova, R., Geradts, Z.: A new model for forensic data extraction from encrypted mobile devices. Digit. Investig. **38**, 301169 (2021)
22. Gandhi, K., Arumugam, C.: Toward a unified and secure approach for extraction of forensic digital evidence from an IoT device. Int. J. Inf. Secur. **22**(2) (2023)
23. Giese, D., Noubir, G.: Amazon echo dot or the reverberating secrets of IoT devices. In: Proceedings of the 14th ACM Conference on Security and Privacy in Wireless and Mobile Networks, pp. 13–24 (2021)
24. Gregorio, J., Alarcos, B., Gardel, A.: Forensic analysis of nucleus RTOS on MTK smartwatches. Digit. Investig. **29**, 55–66 (2019)
25. Grispos, G., Tursi, F., Choo, K.K.R., Mahoney, W., Glisson, W.B.: A digital forensics investigation of a smart scale iot ecosystem. In: 2021 IEEE 20th International Conference on Trust, Security and Privacy in Computing and Communications (TrustCom). pp. 710–717. IEEE (2021)
26. Hadgkiss, M., Morris, S., Paget, S.: Sifting through the ashes: Amazon Fire TV stick acquisition and analysis. Digit. Investig. **28**, 112–118 (2019). https://doi.org/10.1016/j.diin.2019.01.003
27. Hasan, M.: State of IoT 2022: number of connected IoT devices growing 18% to 14.4 billion globally (2022)
28. Horsman, G.: Interpreting digital traces:- 8 foundational pillars to support the formation of opinion in digital forensics. Sci. Justice **64**(1), 38–42 (2024). https://doi.org/10.1016/j.scijus.2023.11.007
29. Hutchinson, S., et al.: Investigating wearable fitness applications: data privacy and digital forensics analysis on android. Appl. Sci. **12**(19), 9747 (2022)
30. Hutchinson, S., Yoon, Y.H., Shantaram, N., Karabiyik, U.: Internet of things forensics in smart homes: design, implementation, and analysis of smart home laboratory. In: 2020 ASEE Virtual Annual Conference Content Access (2020)
31. Iqbal, A., Olegård, J., Ghimire, R., Jamshir, S., Shalaginov, A.: Smart home forensics: an exploratory study on smart plug forensic analysis. In: 2020 IEEE International Conference on Big Data (Big Data), pp. 2283–2290 (2020)
32. Janarthanan, T., Bagheri, M., Zargari, S.: IoT forensics: an overview of the current issues and challenges. Digital Forensic Investigation of Internet of Things (IoT) Devices, pp. 223–254 (2021)
33. Janarthanan, T., Bagheri, M., Zargari, S.: IoT forensics: an overview of the current issues and challenges. Digital Forensic Investigation of Internet of Things (IoT) Devices, pp. 223–254 (2021)
34. Jeon, S., Chung, J., Jeong, D.: Watch out! smartwatches as criminal tool and digital forensic investigations. arXiv preprint arXiv:2308.09092 (2023)
35. Jo, W., et al.: Digital forensic practices and methodologies for AI speaker ecosystems. Digit. Investig. **29**, S80–S93 (2019)
36. Kang, S., Kim, S., Kim, J.: Forensic analysis for IoT fitness trackers and its application. Peer-to-Peer Netw. Appl. **13**, 564–573 (2020)
37. Kaushik, K., Bhardwaj, A., Dahiya, S.: Smart home IoT forensics: current status, challenges, and future directions. In: 2023 International Conference on Advancement in Computation and Computer Technologies (InCACCT), pp. 716–721 (2023)

38. Kim, J., Park, J., Lee, S.: An improved IoT forensic model to identify interconnectivity between things. Forens. Sci. Int.: Digit. Investig. **44**, 301499 (2023)
39. Kim, M., Shin, Y., Jo, W., Shon, T.: Digital forensic analysis of intelligent and smart IoT devices. J. Supercomput. **79**(1), 973–997 (2023)
40. Kim, S., Park, M., Lee, S., Kim, J.: Smart home forensics-data analysis of IoT devices. Electronics **9**(8), 1215 (2020)
41. Krueger, C., McKeown, S.: Using amazon alexa apis as a source of digital evidence. In: 2020 International Conference on Cyber Security and Protection of Digital Services (Cyber Security), pp. 1–8. IEEE (2020)
42. Laricchia, F.: Global smartwatch market share 2020–2022 (2023). https://www.statista.com/statistics/1296818/smartwatch-market-share/. Accessed 09 May 2023
43. Li, S., Choo, K.K.R., Sun, Q., Buchanan, W.J., Cao, J.: Iot forensics: Amazon echo as a use case. IEEE Internet Things J. **6**(4), 6487–6497 (2019)
44. Lin, L., Liu, X., Fu, X., Luo, B., Du, X., Guizani, M.: A non-intrusive method for smart speaker forensics. In: ICC 2021-IEEE International Conference on Communications, pp. 1–6. IEEE (2021)
45. Lorenz, S., Stinehour, S., Chennamaneni, A., Subhani, A., Torre, D.: Iot forensic analysis: a family of experiments with amazon echo devices. Forens. Sci. Int.: Digit. Investig. **45**, 301541 (2023)
46. MacDermott, A., Baker, T., Shi, Q.: Iot forensics: challenges for the IoA era. In: 2018 9th IFIP International Conference on New Technologies, Mobility and Security (NTMS), pp. 1–5. IEEE (2018)
47. MacDermott, A., Lea, S., Iqbal, F., Idowu, I., Shah, B.: Forensic analysis of wearable devices: Fitbit, Garmin and Hetp watches. In: 2019 10th IFIP International Conference on New Technologies, Mobility and Security (NTMS), pp. 1–6. IEEE (2019)
48. Mosenia, A., Jha, N.K.: A comprehensive study of security of internet-of-things. IEEE Trans. Emerg. Top. Comput. **5**(4), 586–602 (2016)
49. Odom, N., Lindmar, J., Hirt, J., Brunty, J.: Forensic inspection of sensitive user data and artifacts from smartwatch wearable devices. J. Forensic Sci. **64**(6), 1673–1686 (2019)
50. Okoli, C., Schabram, K.: A guide to conducting a systematic literature review of information systems research (2010)
51. Oriwoh, E., Jazani, D., Epiphaniou, G., Sant, P.: Internet of things forensics: challenges and approaches. In: 9th IEEE International Conference on Collaborative computing: networking, Applications and Worksharing, pp. 608–615. IEEE (2013)
52. Orr, D.A., Sanchez, L.: Alexa, did you get that? determining the evidentiary value of data stored by the amazon® echo. Digit. Investig. **24** (2018)
53. Pawlaszczyk, D., Friese, J., Hummert, C.: Alexa, tell me-a forensic examination of the amazon echo dot 3rd generation. Int. J. Comput. Sci. Eng. **7**(11), 20–29 (2019)
54. Rongen, J., Geradts, Z.: Extraction and forensic analysis of artifacts on wearables. Int. J. Forens. Sci. Pathol. **5**(1) (2017)
55. Shalaginov, A., Iqbal, A., Olegård, J.: Iot digital forensics readiness in the edge: a roadmap for acquiring digital evidences from intelligent smart applications. In: Edge Computing–EDGE 2020: 4th International Conference, Held as Part of the Services Conference Federation, SCF 2020, Honolulu, 18–20 September 2020, Proceedings 4, pp. 1–17. Springer (2020)
56. Statista. Digital market insights: smart home (2022). https://www.statista.com/outlook/dmo/smart-home/worldwide#revenue. Accessed 09 Apr 2023

57. Stoyanova, M., Nikoloudakis, Y., Panagiotakis, S., Pallis, E., Markakis, E.K.: A survey on the internet of things (IoT) forensics: challenges, approaches, and open issues. IEEE Commun. Surv. Tutor. **22**(2), 1191–1221 (2020)
58. Villarreal, A., Verma, R., Upton, O., Beebe, N.: Non-destructive data acquisition methodology for IoT devices: a case study on amazon echo dot version 2. IEEE Internet of Things J. (2023)
59. Williams, J., MacDermott, A., Stamp, K., Iqbal, F.: Forensic analysis of fitbit versa: Android vs IoS. In: 2021 IEEE Security and Privacy Workshops (SPW) (2021)
60. Xenofontos, C., Zografopoulos, I., Konstantinou, C., Jolfaei, A., Khan, M.K., Choo, K.K.R.: Consumer, commercial, and industrial IoT (in) security: attack taxonomy and case studies. IEEE Internet Things J. **9**(1), 199–221 (2021)
61. Yankson, B., Iqbal, F., Hung, P.: 4p based forensics investigation framework for smart connected toys. In: Proceedings of the 15th International Conference on Availability, Reliability and Security, pp. 1–9 (2020)
62. Yaqoob, I., Hashem, I., Ahmed, A., Kazmi, S., Hong, C.: Internet of things forensics: recent advances, taxonomy, requirements, and open challenges. Futur. Gener. Comput. Syst. **92**, 265–275 (2019)
63. Yildirim, I., Bostanci, E., Güzel, M.S.: Forensic analysis with anti-forensic case studies on amazon Alexa and google assistant build-in smart home speakers. In: 2019 4th International Conference on Computer Science and Engineering (UBMK), pp. 1–3 (2019)
64. Youn, M.A., Lim, Y., Seo, K., Chung, H., Lee, S.: Forensic analysis for ai speaker with display echo show 2nd generation as a case study. Digit. Investig. **38**, 301130 (2021)
65. Zhou, H., Deng, L., Xu, W., Yu, W., Dehlinger, J., Chakraborty, S.: Towards internet of things (IoT) forensics analysis on intelligent robot vacuum systems. In: 2022 IEEE/ACIS 20th International Conference on Software Engineering Research, Management and Applications (SERA), pp. 91–98. IEEE (2022)

Evaluation of Code Similarity Search Strategies in Large-Scale Codebases

Jorge Martinez-Gil[1]($^{\boxtimes}$) and Shaoyi Yin[2]

[1] Software Competence Center Hagenberg GmbH, Softwarepark 32a,
4232 Hagenberg, Austria
`jorge.martinez-gil@scch.at`
[2] Paul Sabatier University, IRIT Laboratory, 118 route de Narbonne, Toulouse,
France
`shaoyi.yin@irit.fr`

Abstract. The ability to automatically identify similar code fragments within huge code repositories is crucial for software development and maintenance tasks such as code reuse and debugging. Although several solutions already exist to face this challenge, not many comparisons have yet been established. For this reason, this study presents a comparative analysis of existing and emerging techniques for code similarity search. We benchmark these methods across diverse codebases, examining metrics such as indexing time, search speed, and the semantic relevance of retrieved code fragments. Our research aims to provide software developers with practical information for performing efficient code similarity searches, addressing the challenges associated with the increasing size of codebases.

Keywords: Code Clone Detection · Code Similarity Search · Codebase Management · Code Reuse · Similarity Search

1 Introduction

Managing large-scale software projects presents several challenges regarding the efficacy and efficiency of the software development process. One significant issue is the potential for duplicated effort [5]. Developers might discover similar code used for comparable tasks as development teams expand and projects increase in complexity. This redundancy wastes valuable time and resources, complicating code maintenance [22]. Additionally, finding relevant code within a large codebase can take much work. Software developers often spend considerable time discovering the specific fragments to modify or extend. These challenges show a need for solutions that maintain consistent code across large-scale projects [21].

Traditional code search methods, such as those using regular expressions, have long been used for searching code fragments within repositories. However, these methods primarily focus on literal string matching, often needing to catch up in capturing the semantic similarity between code fragments. This limitation becomes particularly clear in large codebases where different implementations of similar functionalities may not share exact textual features, so the community widely agrees on the need for more advanced solutions [20].

In recent times, new approaches that generate vector embeddings of code fragments are more appropriate for matching code based on its underlying functionality and not just its textual resemblance. These new approaches allow for more accurate results and faster processing or retrieval of results from large code repositories. Therefore, these techniques provide a novel capability to improve efficiency and decrease redundancy. However, existing solutions still need to be systematically compared. The present study fills this gap, and therefore, the main contributions of this work are as follows:

- An overview of classical and emerging techniques for performing similarity searches in large codebases.
- An empirical evaluation of these techniques with special interest in those relying on semantic code representations rather than literal matching.

The rest of this work is structured fully to ensure a thorough understanding of our research. Section 2 provides a detailed review of existing literature regarding code similarity search. Section 3 presents the methodology usually employed in code similarity search and how we will adapt it to perform our empirical evaluation. In Sect. 4, we perform the experimental setup and present the results regarding the effectiveness and efficiency of existing approaches concerning several benchmarks. Finally, we offer a summary of key findings and potential future directions of this research, providing a reliable roadmap for further exploration.

2 State-of-the-Art

Code similarity search is highly beneficial in several practical scenarios. For example, code reuse involves identifying already-written components, allowing developers to improve efficiency by integrating existing solutions into new projects [14]. In debugging, these techniques can accelerate the process by finding code that has addressed similar bugs or issues in the past, providing a practical reference for developers [11]. Moreover, such techniques facilitate the exploration of different implementations and variations for code understanding, helping developers to identify diverse coding approaches and optimize their solutions [13].

The techniques for code similarity search within large codebases have evolved significantly in recent years by integrating novel techniques [23]. Advanced approaches now go beyond traditional text-based searches to understand the semantic content of code [1]. Techniques like summarization [10] and embedding source code into high-dimensional vector spaces allow systems to assess similarity based on the functionality rather than just textual similarity. Approaches like CodeBERT [4] and GraphCodeBERT [7] are examples of solutions that learn contextual relationships within code, improving the ability to detect similar patterns across diverse codebases. These strategies improve the search accuracy in large repositories, enabling developers to find functionally similar code fragments with different syntactic presentations [11].

Furthermore, the scalability of code similarity search systems has been a critical focus, given the exponential growth of source code in recent years. Systems are now designed to handle vast repositories efficiently, using technologies such as distributed computing, efficient indexing mechanisms, and other tools [12]. One of the most widely used techniques is FAISS [3], which uses a quantization-based approach to compress vectors and speed up the similarity search process without significant losses in precision.

However, more approaches allow for the implementation of near-real-time code search systems [2]. This is because solutions of this kind are crucial for modern software development environments, where developers and even agents need to reuse existing code. These advancements reduce the duplication of effort and contribute to faster development cycles.

Unfortunately, very few empirical studies attempt to rigorously compare the performance of the different emerging solutions for code similarity search. Our work sheds some light on this and provides an overview of what can be expected from these emerging solutions.

3 Problem Statement

Let C_1 and C_2 be source code fragments in two different programming languages, L_1 and L_2, respectively. The code similarity problem involves defining a function $S(C_1, C_2)$ that quantifies their similarity.

Therefore, given $C_1 = \{c_{1,1}, c_{1,2}, \ldots, c_{1,n}\}$ and $C_2 = \{c_{2,1}, c_{2,2}, \ldots, c_{2,m}\}$, where $c_{1,i}$ and $c_{2,j}$ are the atomic elements (such as tokens, statements, abstract syntax tree nodes, etc.) of the code fragments in languages L_1 and L_2, respectively; the similarity function $S(C_1, C_2)$ is a mapping:

$$S : \mathcal{C}_{L_1} \times \mathcal{C}_{L_2} \to [0, 1]$$

where $\mathcal{C}_{L_1}$ and $\mathcal{C}_{L_2}$ are the sets of all possible code fragments L_1 and L_2, respectively, and $S(C_1, C_2) = 1$ indicates maximum similarity and $S(C_1, C_2) = 0$ indicates no similarity.

The function S may be defined based on various criteria, such as syntactic, semantic, or structural likeness between C_1 and C_2, and often involves complex algorithms or machine learning models for its computation [21].

Several effective techniques and tools have been developed to tackle this problem [19]. However, we are only interested in those able to vectorize the code in order to study its scalability with techniques that include Annoy [24], Elasticsearch [6], FAISS [3], HNSW [17], ScaNN [8], and Scikit-Learn-NN [9]. Each technique and tool has some characteristics that suit different scenarios. However, comparing their performance in objective aspects has yet to be studied. Figure 1 illustrates a top-k similarity search in a codebase, focusing on a specific

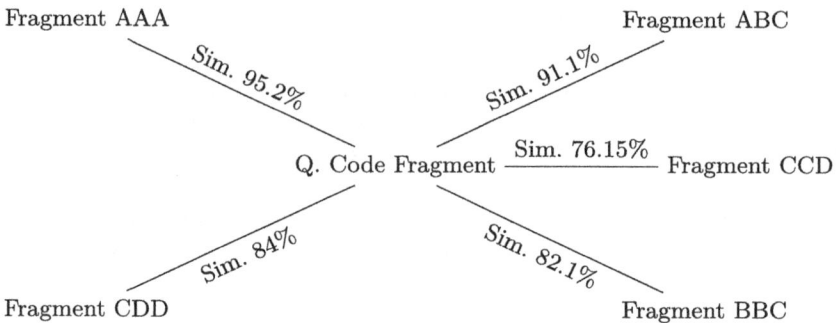

Fig. 1. Example of top-k similarity search in a codebase

query code fragment labeled *Q. Code Fragment*. The diagram uses nodes representing different code fragments and edges to denote the similarity percentage between the query fragment and these fragments.

To date, some of the most outstanding proposals in this area are listed below in alphabetical order:

- Annoy (Approximate Nearest Neighbors Oh Yeah) [24] that performs approximate nearest-neighbor search. It builds forests of trees to partition the space and allows for fast querying, even under very high-dimensional data. Its main advantage is its ability to use memory-mapped files and efficiently handle large-scale datasets.
- Elasticsearch [6] that, although primarily a search engine, can be used for code similarity search through its tunable vector scoring features. It supports text-based and vector-based searches, handling various forms of code and plain language queries. It is particularly advantageous when combining code similarity search with other search capabilities (e.g., text) is essential.
- FAISS (Facebook AI Similarity Search) [3] that is designed for efficient similarity search of high-dimensional vectors. It is beneficial for searching spaces with huge dimensionality, such as embeddings derived from source code. It supports several indexing strategies that optimize speed and accuracy, making it suitable for massive datasets.
- HNSW (Hierarchical Navigable Small World) [17] is an approximate nearest-neighbor search algorithm that uses hierarchical graph structures to efficiently search high-dimensional data. It is known for achieving good recall rates, even in large-scale environments, which is beneficial for code similarity searches where exact matches are not always necessary.
- ScaNN (Scalable Nearest Neighbors) [8] that improves the efficiency of nearest-neighbor computation in high-dimensional spaces using a combination of quantization and tree-based partitioning. It can be tailored to balance accuracy and speed, which is useful when dealing with large codebases.
- SKLNN (Scikit-learn Nearest Neighbors) [9] that supports various algorithms for nearest-neighbor searches, each dealing with different sizes and

dimensionalities. While efficient for small to medium datasets, scalability is assumed to be limited compared to specialized approaches when handling high-dimensional data.

From now on, we will determine the aspects in which these approaches are better suited, which could help give clues as to their use in software development.

4 Main Steps of the Code Similarity Search Process

Modern code similarity search involves several steps, starting with data preprocessing. This includes collecting, cleaning, and normalizing code fragments. Next, these code fragments are transformed into numerical representations through vectorization techniques like TF-IDF [15], or CodeBERT [4]. These vectors are then indexed to allow fast and accurate similarity searches in large code repositories [25]. When a user submits a query, it is also vectorized, and the index is used to retrieve the most similar code fragments efficiently. A critical factor in this search process is k, which determines the number of nearest neighbors returned. Adjusting the k value helps provide results that meet the user's specific needs.

4.1 Data Preprocessing

Effective data preprocessing is crucial for setting up a code similarity search. Initially, code fragment collection can be performed by scraping code repositories, utilizing public dataset compilations, or extracting them from internal project archives. Once collected, code fragments often require cleaning and normalization to ensure consistency and improve effectiveness. This may involve removing comments, normalizing variable names, or standardizing coding styles. Such preprocessing steps help reduce noise and improve the focus on the functional aspects of the code, which are essential for effective similarity searches.

4.2 Vectorization of Code Fragments

Although it is always possible to compare fragment-by-fragment similarity [23], this usually scales poorly. Therefore, vector representations of the fragments are used. Several techniques cover this phase, and the accuracy of the search system depends on the choice. However, since this work focuses on performance rather than accuracy, we will only focus on two techniques: a classical one (TF-IDF) and an emerging one (CodeBERT).

- TF-IDF for representing code fragments presents the ability to assess the importance of terms uniquely relevant to a particular fragment while penalizing standard terms [15]. When working with source code, terms (which could be keywords, function names, or API calls) that appear frequently in a fragment but are rare across other fragments are weighted higher, thus capturing the uniqueness of the fragment. This feature makes TF-IDF particularly suited for distinguishing code fragments that implement specific functionalities, making it a helpful approach for code similarity search [16].

- CodeBERT [4] is a language model designed to process source code automatically. It is trained on a massive dataset of natural language and programming language text, enabling it to perform very well at tasks like code search, completion, and summarization. CodeBERT's bimodal architecture bridges the gap between natural language and code, making it useful for developers who can work with abstract representations of code that are very helpful in a wide range of programming-related tasks.

4.3 Indexing

The need for efficient and scalable similarity searches in large datasets drives the choice of a good indexing strategy. This strategy must be optimized for fast similarity computations over large sets of high-dimensional vectors, making it ideal for handling the vectorized form of code fragments. Then, some kind of distance (e.g., euclidean, cosine, etc.) is used to calculate the similarity between vectors, providing an efficient way to assess the likeness of code fragments based on their vector representations. This setup is particularly effective in environments where fast query responses are crucial, and the dataset size can be huge.

4.4 Similarity Search

During the similarity search process, query code fragments are first vectorized using the same scheme applied during the preprocessing stage. These vectorized forms are fed into an index to find similar code fragments. In this context, the parameter k refers to the number of nearest neighbors considered in the search results. Adjusting k can impact the outcome; a larger k might include more potentially relevant results but also increase the noise, whereas a smaller k focuses on the most similar fragments but may miss some relevant matches. The choice of k thus needs to be balanced based on the specific needs of the search process.

5 Evaluation

In our code similarity search experiments, we have investigated the performance of various techniques and tools under different scenarios. We measured indexing time to assess the efficiency of creating searchable representations of codebases. Accuracy has been evaluated using traditional TF-IDF and the more recent CodeBERT model based on deep learning.

To assess search performance, we have conducted queries on codebases of increasing size: 10,000, 100,000, and 1,000,000 code fragments. We have measured the time taken to retrieve relevant results for each query, providing insights into the scalability of different methods and assessing the efficiency of the different approaches under different workloads.

5.1 Assessment Methodology

A qualitative assessment can provide examples of results showcasing the system's ability to find conceptually similar codes. For instance, a search query for a sorting algorithm could return various implementations of quicksort or mergesort, demonstrating the system's capability to recognize a range of related algorithms. However, this assessment is currently not possible due to the lack of a dataset that has such information annotated, so for the time being, only a quantitative comparison is possible.

Traditionally, code similarity search strategies are compared against simple baselines like keyword searches or basic code representations. However, modern techniques often use vector representations of code, which capture semantic meaning better than keywords. This study compares different vector-based strategies to determine the most effective and efficient way to improve code representation.

5.2 Experimental Setup

In this study, we use the dataset of the BigCloneBench[1] to perform code similarity search. This dataset is a valuable resource for addressing the challenge of identifying duplicate code. Comprising diverse programming languages sourced from real-world software and student projects, it provides a realistic testing ground for evaluating code similarity search strategies. Due to its nature, BigCloneBench is frequently used in solutions to improve code maintainability.

We will always look for the three most similar code fragments. Furthermore, for the parameters of the different strategies, we rely on their default settings, i.e.:

- *Annoy*: Angular distance, 10 trees
- *Elasticsearch*: SVD components 4096, Cosine Similarity
- *FAISS*: FlatL2 index
- *HNSW*: L2 distance, ef_construction: 200, M: 16
- *ScaNN*: L2 normalization, Number of Leaves: 10, Anisotropic Quantization Threshold: 0.2
- *Scikit-Learn NN*: Algorithm K-D Tree

Moreover, the experiments have been performed in an isolated machine on Windows 11 running over an 11th Gen Intel(R) Core(TM) i7-1185G7 at 3.00 GHz. In any case, the reported results are always the result of a relative comparison performed on the same hardware.

---

[1] https://github.com/microsoft/CodeXGLUE/tree/main/Code-Code/Clone-detection-BigCloneBench.

5.3 Performance Comparison

We proceed now to show the experiments performed to establish the performance comparison. These experiments have been designed to evaluate and benchmark the accuracy, and scalability of the different search strategies under various conditions regarding the volume of data to be handled.

Indexing Time. Figure 2 presents a comparative summary of the time required to index code fragments using various approaches for different dataset sizes: 10k, 100k, and 1M code fragments. The table provides an overview of the performance of the six different indexing approaches under consideration in this study.

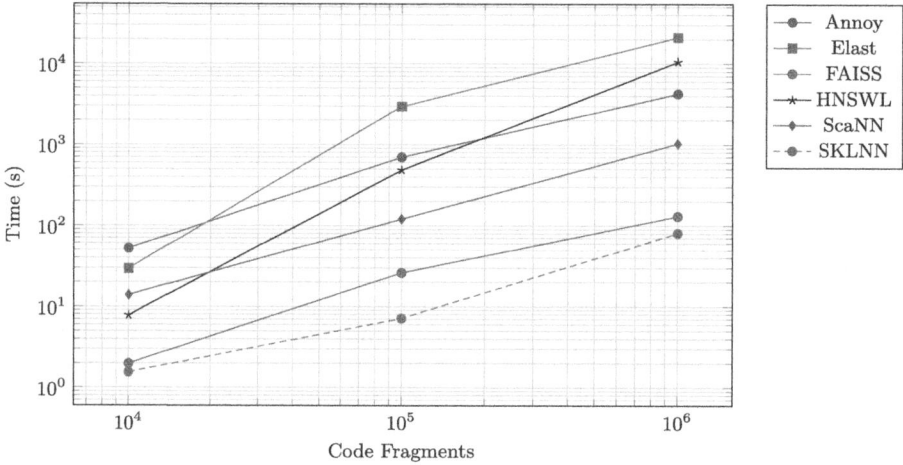

Fig. 2. Performance comparison of different methods for 10k, 100k, and 1M code fragments

All techniques show an expected increase in indexing time as the dataset size grows. The values for each technique suggest varying degrees of scalability. However, FAISS, which shows a relatively smaller increase in time, might indicate better scalability than others in the long run. This is deduced from the lower slope inclination, which will cause it to increase at a lower rate than the rest.

It is also necessary to note that all indexes (except Elasticsearch) are kept in the main memory. Only Elasticsearch, as a complete DBMS, works directly on optimized disk files and is consolidated in secondary memory.

Accuracy. Code similarity search aims to find duplicate or similar code fragments but involves a trade-off between accuracy and efficiency. The chosen algorithm and parameters can affect the results. Exact matching ensures high precision but may overlook functionally similar code with different structures. In contrast, approximate methods like token-based or semantic analysis cover more ground but can produce false positives. The quality and relevance of training data also affect the accuracy of machine learning-based approaches.

Figure 3 illustrates the relationship between the number of code fragments and accuracy using TF-IDF over FAISS. The x-axis represents the number of code fragments on a logarithmic scale, and the y-axis represents the accuracy. The figure shows that the accuracy tends to decrease as the number of code fragments increases. This downward trend suggests a negative correlation between the number of code fragments and the accuracy in this context.

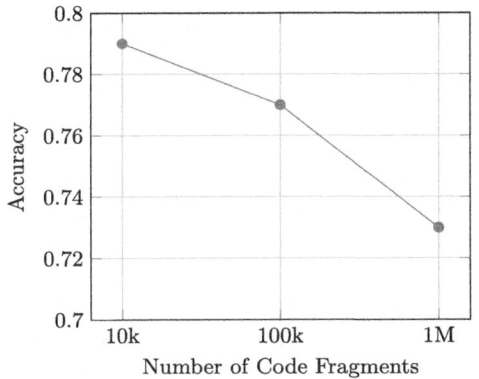

Fig. 3. Search accuracy using TF-IDF

Figure 4 illustrates the relationship between the number of code fragments and accuracy using CodeBERT over FAISS. The x-axis represents the number of code fragments on a logarithmic scale, and the y-axis represents the accuracy. The figure shows that the accuracy remains relatively stable as the number of code fragments increases. There is a slight decrease in accuracy when the number of code fragments reaches 1,000,000, but overall, the accuracy stays high, indicating a good performance across different scales.

However, this work does not seek to compare semantic similarity models. Previous studies already exist and even show that some variants of CodeBERT [7] and an ensemble of simple models for assessing semantic similarity tend to perform well [18].

Search Performance. Search performance refers to how fast a given approach can compare source code fragments. Faster performance is crucial for real-time or large-scale applications. We have studied this performance for 10k, 100k, and 1M code fragments. Times were measured five times, and the median value was reported.

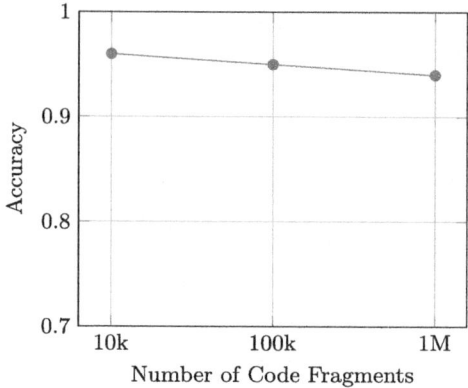

Fig. 4. Search accuracy using CodeBERT

Experiment 1: 10,000 Items. Figure 5 shows us that the query performance of the different approaches varies significantly. The comparison shows how some approaches maintain lower search times consistently, which is critical for applications requiring efficient code retrieval. This is important to avoid potential bottlenecks in systems where fast code retrieval is essential, such as in integrated development environments or real-time code analysis tools.

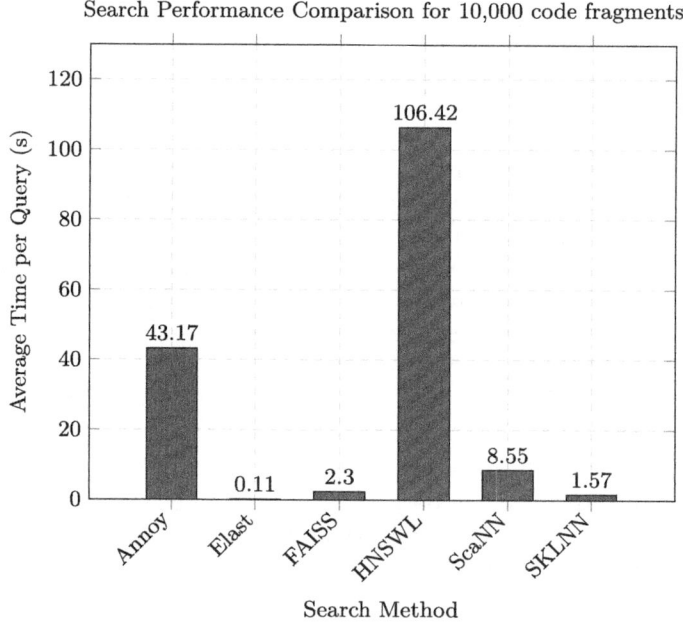

Fig. 5. Search performance for a codebase of 10,000 items

Experiment 2: 100,000 Items. Figure 6 shows us that the query performance of different approaches varies again. The performance trend observed with 10,000 items continues here, with specific approaches (Elasticsearch, SKLNN, FAISS) demonstrating better scalability as the dataset size increases. So, they could be good candidates when choosing the right strategy for large-scale data, assuring that query performance remains efficient even as data volume grows. Therefore, we can see which methods could be suitable for handling expansive datasets to guarantee reliable performance in large-scale applications.

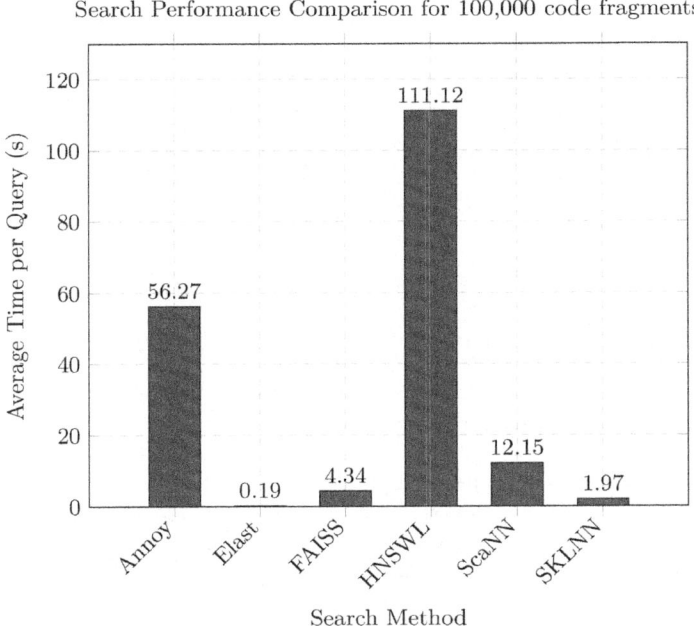

Fig. 6. Search performance for a codebase of 100,000 items

Experiment 3: 1,000,000 Items. Figure 7 shows us again that the query performance of the different approaches varies a lot. This figure shows us the importance of choosing the right approach for large-scale applications, as the difference in search times can be substantial. Therefore, it seems clear that selecting an efficient method is crucial for maintaining stable performance in systems with large datasets. The rationale behind opting for the right solution is to improve system responsiveness and user satisfaction even as the code volume grows.

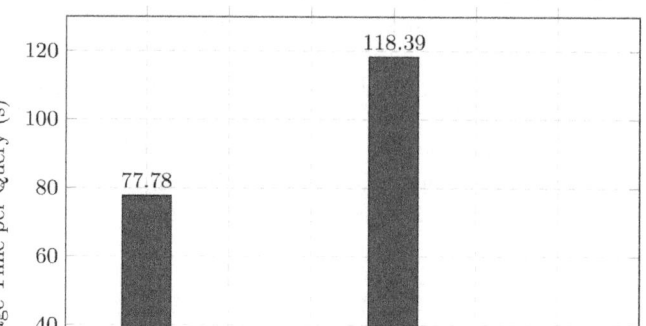

Fig. 7. Search performance for a codebase of 1,000,000 items

5.4 Summary of Results

The results of our experiments reveal some significant findings that could be interesting when guiding developers and researchers in selecting the most suitable strategies for their specific use cases and requirements regarding code similarity search. The following facts can be extracted from these results:

- CodeBERT is the best strategy to maximize accuracy. These results confirm the findings that have also been obtained in the frame of several previous research studies [7,19].
- Elasticsearch produces the best results in terms of search performance but at the cost of having the worst indexing time.
- The approach with the best scalability prospects, in the long run, is FAISS, and in the short term, is SKLNN.
- Most approaches present good performance, making them viable candidates for integration into production systems, although Elasticsearch is recommended to consolidate the index in secondary memory.
- The results of SKLNN are positively surprising since it is part of the general-purpose Scikit-learn library, which is very popular in the software development world. In fact, for small samples, it has very good indexing and search speed times.

6 Discussion

Code similarity search helps developers identify algorithm variations, aiding in code understanding and optimization. Identifying similar code fragments across different projects or within the same codebase facilitates developers' efficient search for duplicated code fragments, which is crucial for improving their projects' maintainability.

The process of vectorizing code enables code similarity search to effectively capture the essence of the code, regardless of varying structures and styles. Vectorization transforms code into a numerical representation, allowing the different strategies for code similarity search to analyze and compare code fragments more effectively.

However, despite their advantages, code similarity search strategies face challenges, such as complex syntax and dependencies in programming languages. Effective vectorization must capture both syntactic and semantic aspects of code, which can be difficult due to varying structures and styles. Furthermore, vocabulary mismatch is another issue. Different naming conventions and synonyms can lead to inaccurate search results.

We have seen that as codebases grow, search performance can degrade. Efficient indexing and retrieval mechanisms are necessary to maintain high performance in large-scale applications. Addressing these limitations is crucial to improving code similarity search. Advances in this context offer promise in overcoming some challenges by developing models that better understand code context and semantics. Standardized naming conventions and coding practices could also reduce vocabulary mismatches and improve search accuracy.

7 Conclusion

This study has evaluated existing code similarity search techniques, focusing on their effectiveness and scalability in real-world scenarios. Our benchmark results demonstrate that the studied methods are suitable for performing code similarity searches, although with some objections. The key idea is capturing the functional essence of the code beyond mere textual similarity, improving the accuracy of search results within large codebases. This approach has proven invaluable for developers, enabling them to efficiently identify functionally similar but syntactically diverse code fragments.

We have established a comparative analysis to identify the best existing approaches to working in a context where codebases grow rapidly. Our experiments show that most approaches offer good average performance. However, some are much better than the rest in some aspects related to the results indexation, search speed, and semantic relevance.

Furthermore, we have identified several areas for future research. Developing more efficient indexing techniques could improve the scalability of code similarity search, particularly for large codebases. Exploring novel approaches to capture the semantic meaning of code could improve the relevance of retrieved code fragments and optimize the search process.

Acknowledgments. The authors thank the anonymous reviewers for their help in improving the work. The research reported in this paper has been funded by the Federal Ministry for Climate Action, Environment, Energy, Mobility, Innovation, and Technology (BMK), the Federal Ministry for Digital and Economic Affairs (BMDW), and the State of Upper Austria in the frame of SCCH, a center in the COMET - Competence Centers for Excellent Technologies Programme managed by Austrian Research Promotion Agency FFG.

References

1. Ain, Q.U., Butt, W.H., Anwar, M.W., Azam, F., Maqbool, B.: A systematic review on code clone detection. IEEE Access **7**, 86121–86144 (2019)
2. Aumüller, M., Bernhardsson, E., Faithfull, A.: ANN-benchmarks: a benchmarking tool for approximate nearest neighbor algorithms. Inf. Syst. **87**, 101374 (2020)
3. Douze, M., et al.: The Faiss library (2024)
4. Feng, Z., et al.: CodeBERT: a pre-trained model for programming and natural languages. In: Cohn, T., He, Y., Liu, Y. (eds.) Findings of the Association for Computational Linguistics: EMNLP 2020, Online Event, 16–20 November 2020, volume EMNLP 2020 of *Findings of ACL*, pp. 1536–1547. Association for Computational Linguistics (2020)
5. Gabel, M., Jiang, L., Su, Z.: Scalable detection of semantic clones. In: Proceedings of the 30th International Conference on Software Engineering, pp. 321–330 (2008)
6. Gormley, C., Tong, Z.: Elasticsearch: The Definitive Guide: A Distributed Real-Time Search and Analytics Engine. O'Reilly Media Inc, Sebastopol (2015)
7. Guo, D., et al.: GraphcodeBERT: Pre-training code representations with data flow. arXiv preprint arXiv:2009.08366 (2020)
8. Guo, R., et al.: Accelerating large-scale inference with anisotropic vector quantization. In: International Conference on Machine Learning, pp. 3887–3896. PMLR (2020)
9. Hackeling, G.: Mastering Machine Learning with Scikit-Learn. Packt Publishing Ltd, Birmingham (2017)
10. Haque, S., Eberhart, Z., Bansal, A., McMillan, C.: Semantic similarity metrics for evaluating source code summarization. In: Proceedings of the 30th IEEE/ACM International Conference on Program Comprehension, pp. 36–47 (2022)
11. Higo, Y., Ueda, Y., Kamiya, T., Kusumoto, S., Inoue, K.: On software maintenance process improvement based on code clone analysis. In: Oivo, M., Komi-Sirviö, S. (eds.) PROFES 2002. LNCS, vol. 2559, pp. 185–197. Springer, Heidelberg (2002). https://doi.org/10.1007/3-540-36209-6_17
12. Inoue, K., Miyamoto, Y., German, D.M., Ishio, T.: Finding code-clone snippets in large source-code collection by ccgrep. In: Taibi, D., Lenarduzzi, V., Kilamo, T., Zacchiroli, S. (eds.) OSS 2021. IAICT, vol. 624, pp. 28–41. Springer, Cham (2021). https://doi.org/10.1007/978-3-030-75251-4_3
13. Juergens, E., Deissenboeck, F., Hummel, B., Wagner, S.: Do code clones matter? In: 2009 IEEE 31st International Conference on Software Engineering, pp. 485–495. IEEE (2009)
14. Karmakar, A., Robbes, R.: What do pre-trained code models know about code? In: 2021 36th IEEE/ACM International Conference on Automated Software Engineering (ASE), pp. 1332–1336. IEEE (2021)
15. Karnalim, O.: TF-IDF inspired detection for cross-language source code plagiarism and collusion. Comput. Sci. **21** (2020)

16. Karnalim, O., et al.: Explanation in code similarity investigation. IEEE Access **9**, 59935–59948 (2021)
17. Malkov, Y.A., Yashunin, D.A.: Efficient and robust approximate nearest neighbor search using hierarchical navigable small world graphs. IEEE Trans. Pattern Anal. Mach. Intell. **42**(4), 824–836 (2018)
18. Martinez-Gil, J.: A comprehensive review of stacking methods for semantic similarity measurement. Mach. Learn. Appl. **10**, 100423 (2022)
19. Martinez-Gil, J.: Source code clone detection using unsupervised similarity measures. In: Bludau, P., Ramler, R., Winkler, D., Bergsmann, J. (eds.) SWQD 2024. LNBIP, vol. 505, pp. 21–37. Springer, Cham (2024). https://doi.org/10.1007/978-3-031-56281-5_2
20. Novak, M., Joy, M., Kermek, D.: Source-code similarity detection and detection tools used in academia: a systematic review. ACM Trans. Comput. Educ. (TOCE) **19**(3), 1–37 (2019)
21. Roy, C.K., Cordy, J.R., Koschke, R.: Comparison and evaluation of code clone detection techniques and tools: a qualitative approach. Sci. Comput. Program. **74**(7), 470–495 (2009)
22. Saini, N., Singh, S., et al.: Code clones: detection and management. Procedia Comput. Sci. **132**, 718–727 (2018)
23. Satter, A., Sakib, K.: A similarity-based method retrieval technique to improve effectiveness in code search. In: Companion Proceedings of the 1st International Conference on the Art, Science, and Engineering of Programming, pp. 1–3 (2017)
24. Spotify. Annoy. https://github.com/spotify/annoy, May 2023. Approximate Nearest Neighbors in C++/Python optimized for memory usage and loading/saving to disk
25. Tronícek, Z.: Indexing source code and clone detection. Inf. Softw. Technol. **144**, 106805 (2022)

Quality Assessment of Volunteered Geographic Information: A Survey

Donia Nciri[1(✉)], Salma Sassi[2], Richard Chbeir[2], and Sami Faiz[3]

[1] Tunisia Polytechnic School, La Marsa, Tunisia
donia.nciri@ept.u-carthage.tn
[2] Univ. Pau et Pays Adour UPPA, LIUPPA, Anglet, France
{salma.sassi,richard.chbeir}@univ-pau.fr
[3] Higher Institute of Multimedia and Arts of Manouba, La Manouba, Tunisia
sami.faiz@insat.rnu.tn

Abstract. Traditionally, government and national mapping agencies have been a primary provider of authoritative geospatial information. Today, with the exponential proliferation of Information and Communication Technologies or ICTs (such as GPS, mobile mapping and geo-localized web applications, social media), any user becomes able to produce geospatial information. This participatory production of geographical data gives birth to the concept of Volunteered Geographic Information (VGI). This phenomenon has greatly contributed to the production of huge amounts of heterogeneous data (structured data, textual documents, images, videos, etc.). It has emerged as a potential source of geographic information in many application areas. Despite the various advantages associated with it, this information lacks often quality assurance, since it is provided by diverse user profiles. To address this issue, numerous research studies have been proposed to assess VGI quality in order to help extract relevant content. This work attempts to provide an overall review of VGI quality assessment methods over the last decade. It also investigates varied quality assessment attributes adopted in recent works. Moreover, it presents a classification that forms a basis for future research. Finally, it discusses in detail the relevance and the main limitations of existing approaches and outlines some guidelines for future developments.

Keywords: VGI · Quality assessment · Spatial Data quality · Quality measures and indicators

1 Introduction

Nowadays, with the evolution of Information and Communication Technologies (ICTs), and the arrival of web 4.0 and supporting technologies, users can produce

---

S. Sassi, R. Chbeir and S. Faiz—These authors contributed equally to this work.

more and more geographical information from their own and contribute to the creation of massive content and dissemination of geographic data through editing online maps, publication of geo-referenced images or geotags. This allows connecting with information and giving rise to the concept of Volunteered Geographic Information (VGI). The term VGI was introduced by authors in reference [1] as the widespread participation of a large number of citizens, often with little official competence, in the creation of geographic information. Indeed, web-mapping services are a form of digital mapping based on the use of the internet to produce, design, process and publish maps. Since the adoption of Web 2.0, many online mapping services have emerged, such as Google Maps[1], Google Earth[2], Bing Maps[3], and OpenStreetMap[4](OSM) based on cooperative approaches. VGI can take different forms (as indicated in Chap. 2): geotagged photographs through sites such as Panoramio and Flickr, online maps such as OpenStreetMap (OSM) and Wikimapia, or 3D VGI such as OSM-3D and OSM2World [2]. Also, the concept of volunteered information is often related to information produced using social networks, generally considered as the most popular sources for collecting volunteered data, since they allow hundreds of millions of Internet users in the whole world to produce and consume the generated content. Recently, VGI has been used as a useful tool for crisis management [3]. For example, it allows victims of natural disasters (such as forest fires or epidemics), to connect quickly with the rest of the world mainly through sharing photos and videos, in order to minimize the effects of disasters and in emergency response assistance [4]. It also provides a new channel of communication for government agencies to reach relevant citizen information [5].

Although the collection of VGI offers major advantages for the study of geography, it also presents serious threats related to the veracity and accuracy of data. In addition, unlike most traditional sources, VGI may not be subjected to filtering by professional quality controllers, and it often lacks identity indicators such as the author or the source. Similarly, there are no universal standards for posting information online. Numerical data can be easily modified, plagiarized, distorted, or created anonymously under false pretexts [6]. VGI quality analysis and assessment have become a very challenging issue amongst academics and researchers [7] who have investigated several approaches such as [8–11], and [12]. These approaches mainly differ with regard to the type of information evaluated and the reference data types, among other factors. Several studies have reviewed and categorized existing approaches such as [3,7,13–16] and [17]. However, these studies do not differentiate clearly between the methods used for each type of VGI and the quality attributes associated to each method. Moreover, each existing survey deals with a specific type of VGI and specific application area. None of them reviews VGI quality methods and attributes for all VGI types that cover all existing domains.

---

[1] http://www.google.com/maps.
[2] https://www.google.com/earth/.
[3] https://www.bing.com/maps/.
[4] https://openstreetmap.org/.

This article can be regarded as an extension of previous studies which aims at proposing an overall overview for assessing VGI quality for both methods and quality attributes. Over recent years, a significant number of researchers have already presented some works about the quality of data in GIS and VGI, including literature reviews. For example, an overview of measures and indicators of VGI quality is carried out by the authors in reference [7]. Then, the authors classify these indicators in categories according to data, demographics, socio-economic situation and contributors. This review provides the basis for the ongoing academic effort to create a practical quality evaluation method through the use of appropriate quality indicators. The authors in reference [18] presented an overview of the strategies currently adopted to improve VGI quality is executed. The paper reviews and categorizes the VGI in Citizen Science projects. The authors in reference [3] proposed a review to discover current state of knowledge on Crowdsourced Geographic Information (CGI) and its relationship with disaster management and presents recommendations for future research. The research investigates how VGI may present an opportunity to connect and engage individuals in disaster events. Furthermore, reference [14] analyzed the main quality attributes that can be applied in the VGI context and identifies, in the literature, methods that contribute to increasing VGI quality. Also, this survey proposes new methods that can help obtain data with quality assurance. The authors in reference [13] reviewed various quality measures and indicators for selected types of VGI, and existing quality assessment methods. That research work presents a classification of VGI with current methods utilized to assess the quality of Map based VGI.

Also, the authors in reference [15] presented the results of Systematic Literature Review (SLR) that is carried out to discover the methods that can be employed to assess the quality of CGI in the absence of authoritative data. For the accomplishment of this work, the authors of reference [16] proposed a taxonomy of methods for assessing the quality of VGI when no reference data is available. These methods have been identified by means of previous systematic literature review. One more recent SLR proposed by the authors in reference [17] that seeks to analyze data quality in GIS and VGI. The purpose of this paper is to provide an overview of a set of papers extracted from a list of prestigious scientific libraries and verify the quality of data present in these types of system. Table 1 presents a list of existing surveys.

Several survey papers of VGI quality assessment were conducted to provide an overview and allow researchers working on this area. As shown in Table 1, some studies include methods to evaluate VGI quality [3,15,18], and [16], while, others only include measure for VGI quality assessment [7]. However, some authors in references [3] and [17] presented an overview of both but they are limited in map-based VGI. Fewer efforts have looked at the quality of text-based and image-based VGI. In addition, some surveys papers invoked existing methods and measures without putting them in a taxonomy. In fact, classifying and categorizing existing methods is an essential step for future work since it can help authors to extend existing approaches or combine them to develop new

Table 1. A list of Existing Surveys

Paper	Year	VGI Methods	VGI Measures	Classification	Paper Objective
[18]	2014	Yes	No	Yes	Citizen Science project
[7]	2015	No	Yes	Yes	Map-based VGI
[3]	2015	Yes	No	No	Disaster management
[14]	2016	Yes	Yes	Yes	Map-based VGI
[13]	2017	Yes	Yes	Yes	Map-based VGI
[15]	2017	Yes	No	Yes	No reference data
[16]	2018	Yes	No	Yes	No reference data
[17]	2019	Yes	Yes	No	Map-based VGI

methods in various application areas. Indeed, no studies were found that include the classification of methods and measures for VGI quality assessment in various sources (e.g., social media, crowdsensing, collaborative mapping). Given the large number of researches carried out to evaluate VGI quality, an update is necessary to identify the new methods and measures in the literature. Our survey comes to analyze and categorize them in detail taking into account all VGI sources and types. In other terms, our study addresses and answers the following questions: what is the source of VGI in which each method could be employed? What quality attributes are associated to each method? What methods can be used to assess the quality of VGI when authoritative data is not available? And what is the field of application of each method? This study analyzes the recent methods (over the period 2011–2021) that contribute to increasing VGI quality, and identify the main quality attributes that can be applied to assess VGI quality.

The remainder of this review is organized as follows. Section 2 presents the adopted literature review methodology. Section 3 presents the main concepts of VGI quality. Section 4 reviews existing methods and quality attributes for VGI quality assessment. In Sect. 5, we discuss the limitations and future research directions and perspectives. Last Sect. 6 is dedicated to conclude this study.

2 The Literature Review Methodology

The main objectives of this study are to study, understand, summarize and provide an overview of existing approaches and tools used for assessing VGI quality. To explore all the existing studies discussed in literature, we used a rigorous review protocol based on the strategy of Systematic Literature Review (SLR) detailed in the next section.

2.1 Research Strategies

The study carried out using the "Publish Or Perish" (POP) software. It is a free software program that analyzes academic citations based on data collected by the specialized indexes such as Google Scholar, Microsoft Academic's, Scopus, PubMed, Web of Science, to mention a few.

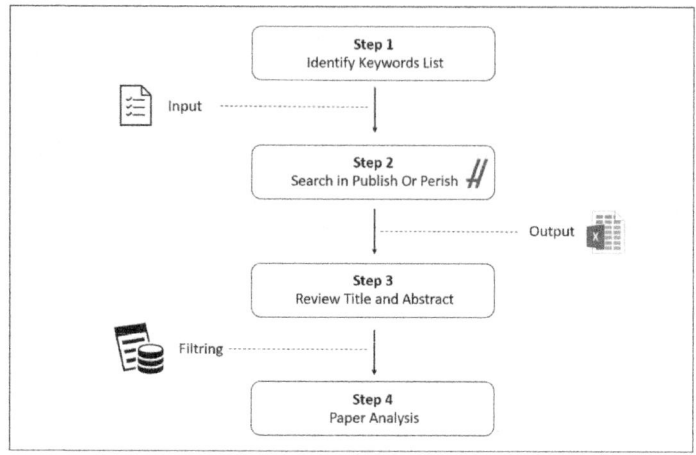

Fig. 1. Methodology Process

As shown in Fig. 1, the methodology that we adopted was based on three main steps. As a first step, we began by performing a search based on a predefined set of keywords (e.g., VGI, Quality assessment). The objective was to retrieve all the terms, words and expressions related to our subjects. We also identified synonyms for each term in order to maximize the number of returned studies. For example, in some studies, the term "VGI: Volunteered Geographic Information" was replaced by the term "CGI: Crowdsourced Geographic Information". Based on those keywords we carried out our research using the following search string ((("VGI" OR "CGI") data quality") AND ("quality dimensions" OR "quality measures" OR "quality indicators" OR "quality metrics) AND ("quality assessment" OR "quality evaluation").

Table 2. The inclusion and exclusion criteria

Inclusion criteria	Exclusion criteria
IC1: studies published between 2010 and 2023	EC1: studies that are duplicates
IC2: studies published in English	EC2: studies that are not accessible online
IC3: studies focused on VGI data quality assessment	EC3: Master's and doctoral dissertations

Next, as a second step, we used the corresponding queries of this search string as input in POP software to perform our search. The search was realized on the

Quality Assessment of Volunteered Geographic Information: A Survey 119

following scientific libraries and electronic databases that index the main journals and conferences of the area: Google Scholar search engine, web of science, ACM Digital Library, Science Direct, and Scopus. Figure 2 shows the result returned by the POP tool for the query "Crowdsourced Geographic Information quality".

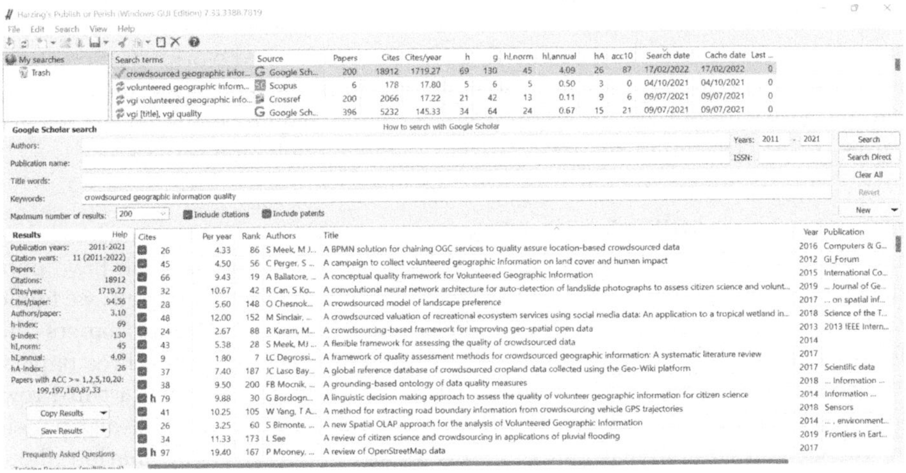

Fig. 2. List of selected paper generated by POP research software

Then, we exported the result of each query to a worksheet of an excel file by including several meta-data provided by POP such as the name of author(s), the type of paper, the publication year, the name of publisher, etc. Figure 3 shows the PRISMA flow diagram.

Afterwards, we merged all the worksheets into one final worksheet and obtained more than 200 papers[5]. After removing duplicates and reviewing title and abstract, we selected 65 relevant papers written in English[6]. Our selection was based on studies published between 2010 and 2023 that proposed new methods for data quality assessment for VGI. Table 2 provides a description of the inclusion and exclusion criteria used in our review and accordingly we decide whether or not to select each study for further analysis.

2.2 Search Result Analysis

2.2.1 Publication Years and Source Overview

As described previously, the literature search was carried out using two different strategies; automatic and manual. The automatic search was performed using

---

[5] https://www.dropbox.com/s/3p6u3x7gij2wadx/POP%20Result%20fo%20VGI.xlsx?dl=0.

[6] https://www.dropbox.com/s/4dipl0tlc1tim4r/POP%20Result%20selected.xlsx?dl=0.

POP software in order to obtain the list of existing studies. Hence the manual search was made to narrow down the research. After analyzing obtained result, we observe that the distribution of relevant papers over the selected period show that most research studies that focused on VGI quality assessment are obtained in the period between 2013 and 2018. In last years, few efforts have been made to assess VGI quality. These results can be observed in Fig. 4. Then, the majority of the studies were published and cited in journals and conferences. Only five research concerns book chapter and there is none of studied works that concerns thesis research. Figure 5 presents the distribution of these studies derived from their publication sources. It shows that reviewed papers included 42 journal articles (64%), 18 conference reports (28%), and 5 book chapters (8%).

2.2.2 Authors Affiliation Overview

Analysis of the obtained result identified many active authors which have published more than one studies in book, journal or conference. In the front, we found "Zipf. A" with five papers. In the second rank, "See. L" with four papers and then, "Devillers. R", "Fritz. S", "Foody. G" and "Perger. C" with three papers. Others such as "Mooney. P", "Comber. A", and "Forati. A" come in third row. Table 3 list the most highly cited authors with the number of published papers and the type of the publication. In addition, after analyzing information about authors and their affiliations, we observed that most selected paper belongs to University of Applied Sciences (45%). These affiliations are derived from Austria. Heidelberg University (Germany) is the second affiliation with (25%). Then,

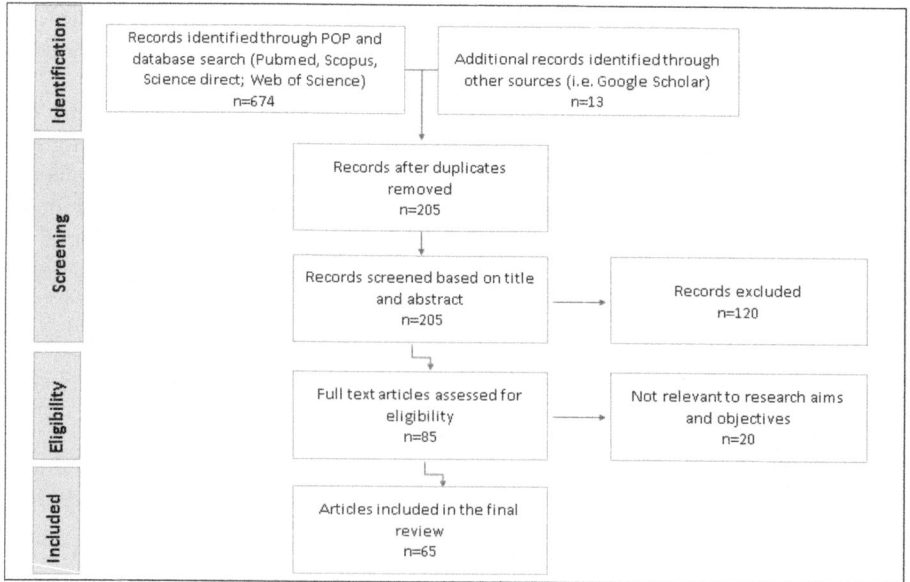

Fig. 3. The PRISMA flow diagram

we found the Institute of Research for Development (France) in the third place. Others university such as Maynooth University (Ireland) and University of Nottingham (UK) also present a significant number of research in the field. Figure 6 describes the percentage of each affiliation in further details.

3   VGI Quality Concept and Foundations

Quality is an old and recurring concern that remains crucial in several fields other than geomatics. Indeed, different meanings have been associated with the concept of quality, mainly from the sector of the manufacture and distribution of products and services. The International Organization for Standardization (ISO)[7] defines data quality as "adequacy to requirements; meeting the needs of the user" and the quality of a product as "the totality of the characteristics of a

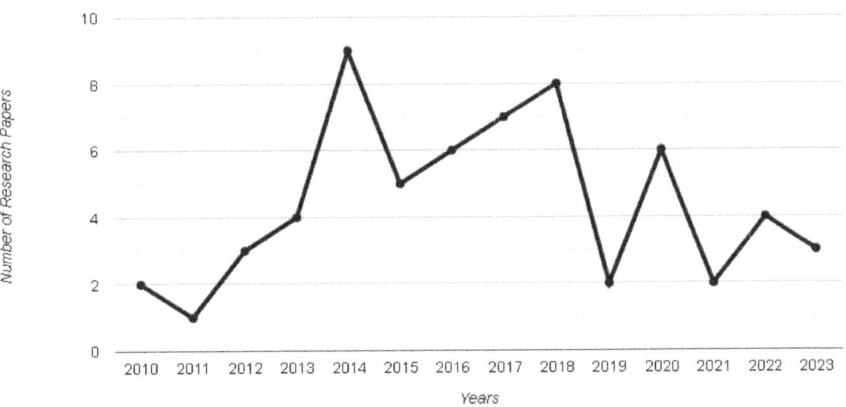

Fig. 4. Distribution of papers over the period 2010–2023

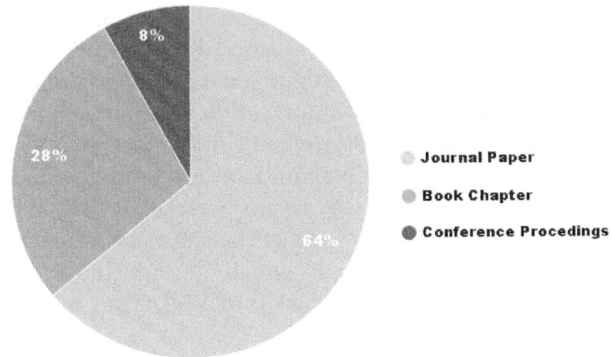

Fig. 5. Publication source overview

---

[7] http://www.iso.org.

Table 3. Most active authors and corresponding computed publications

Author's name	Number of research	Type of publication
Zipf. A	5	Journal, Conference
See. L	4	Book chapter, Journal
Devillers. R	3	Book chapter, Journal
Fritz. S	3	Book chapter, Journal
Perger. C	3	Journal
Foody. G	3	Book chapter, Journal
Mooney. P	2	Book chapter, Journal, Conference
Karimipour. F	2	Journal, Conference
Comber. A	2	Journal
Vandecasteele. A	2	Book chapter, Journal
Barron. C	2	Journal, Conference
Schill. C	2	Journal
Eckle. M	2	Conference
Vahidi. H	2	Journal, Conference
Chehreghan. A	2	Journal
Ali Abbaspour. R	2	Journal
Jabeur. N	2	Journal, Conference
Forati. A	2	Conference

product or service that affect its ability to satisfy the explicit or implicit needs of the customer". For some authors (such as in [19]), the definition of information quality is relative to the user. Others such as [20] suppose that data quality definition and assessment in Big Data mainly depend on the data types, data sources and applications. In [1], the authors mentioned that data quality assurance is done in two parts: the first by assuring and checking data quality during the contribution and creating them, the second by checking data quality after making them by a comparison with reference data and save the results in form of metadata. Volunteered geographic information is a particular type of crowdsourced data that comes from a wide range of different sources, and its quality has also a great importance since it should be assessed specifically according to the source and the type of analysis to be performed [21]. Therefore, many factors can affect the quality of VGI, such as the characteristics of the volunteer, the type of information, and the way in which the information is produced as discussed in [18]. An interesting classification of VGI types is introduced in [22] such as:

- Social media: concerns geographic information shared in social media,
- Crowd sensing: includes the use of collaborative technologies for gathering observations by means of specific platforms,

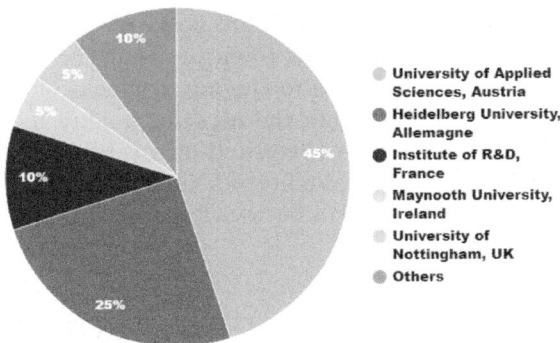

Fig. 6. Author Affiliation and Countries

Table 4. ISO measurement

ISO measurement	Description
Completeness	Represents the proportion of stored data against the potential of 100% complete [24]
Logical consistency	Concerns the consistency of different database objects with other objects of the same theme (intra-theme consistency) or objects of other themes (inter-theme consistency) [25]
Positional accuracy	Analyses the accuracy of the position of features within a spatial reference system. In other terms, it represents the positional difference between two geospatial features or between a geospatial features and reality. [26]
Thematic accuracy	calculates the degree to which the attributes of a map agree with ground-truth dataset [27]
Temporal quality	Refers to the evolution of VGI data in time [7]
Usability	Analyses the usefulness and ability of the data to meet user needs [28]

- Collaborative mapping: requires volunteers to produce a very specific type of georeferenced data, for instance geographic data about points of interest, streets, roads, buildings, land use, etc.

ISO/TC 2117 (Technical Committee)[8] proposed a set of international standards that define the measures of geographic data quality. These measures are: *completeness, logical consistency, positional accuracy, temporal quality, thematic accuracy*, and *usability*. In the literature, there is a set of methods for assess-

---

[8] https://www.iso.org/technical-committees.html.

ing the quality of geographic data, which are generally based on the theory of measurement. However, it is important to ensure that the results of the quality assessment are valid to contribute to the improvement of disseminated geographic information quality. Although the need for quality measures and for a quality framework for VGI has been advocated by several authors [2,7] and [23], neither a shared definition of VGI quality nor indicators for such quality have been clearly established. Table 4 introduce ISO quality measures for geographic data quality.

4 Quality Methods and Measures for VGI

In this section, we introduce the taxonomy resulting from the analysis of the papers selected by our review. At first, we present a detailed description of the existing quality dimensions proposed for the issue of VGI data quality. Then, we discuss existing methods and approaches for assessing the quality of VGI.

4.1 VGI Quality Measurement

The quality of VGI can vary depending on factors such as the motivation of the contributors, their expertise in a specific domain, and the tools used to collect and validate the data. To ensure the quality of VGI, there are various methods, techniques and measurement. VGI quality measurement refers to the set of metrics that are used to assess the quality of geographic information collected and provided by volunteers. Ultimately, the objective of VGI quality measurement is to support the development of a robust and reliable source of geographic information that can be used for a variety of purposes. Evaluating the quality of VGI and recovering its hidden knowledge still a great challenge [29] and [26]. Assessing VGI quality may be analogous to assessing the content of websites since VGI is the content of systems that collect and distribute georeferenced volunteered information [14]. Several studies have been proposed (as discussed in the following subsection) and are generally based on a set of basic criteria that aim at evaluating the credibility of geographic information [30], mainly, ISO quality measurement for geographic information: completeness, positional accuracy, thematic accuracy, logical consistency, temporal quality and usability).

To measure VGI quality, additional specific features are needed to conduct a quality assessment different from the case of traditional geographic data [31]. According to [1], there are three kinds of measures to ensure VGI data quality. The first is crowd-sourcing revision, where data quality can be ensured by multiple contributors. The second, social measures, which focuses on the assessment of contributors themselves as a proxy measure for the quality of their contributions; and the third, geographic consistency, that allow an analysis of the consistency of contributed entities. In [18], the authors classify VGI quality indicators into two main categories: internal and external quality. The internal quality measures are grouped by type of VGI, i.e. measurements or text-based. While, the external quality measures are grouped by reliability of the individual and reputation of the organization. Others, such as [16], suggest a taxonomy based on two mainly characteristics: 1) extrinsic quality when the criteria use external knowledge, and, 2) intrinsic quality when the criteria do not use external knowledge.

Various research initiatives have been carried out to propose new indicators for assessing VGI quality. For example, in [32], the authors use plausibility as a measure to predict the correct class of a new entity in OpenStreetMap. As well, the authors in [33] and [34] implement trustworthiness to ensure VGI quality produced by participatory sensing and collaborative mapping. In the same context, the authors of [35] allow analyzing volunteer profiles to determine the trustworthiness of his/her posts on social media.

In this study, we review existing criteria that were used to evaluate VGI quality. Following this review, we have found 17 quality measures and indicators that addressed within the 57 papers we surveyed which are: Positional Accuracy, Thematic Accuracy, Completeness, Logical Consistency, Reputation, Credibility, Trust, Reliability, Trustworthiness, Semantic Accuracy, Temporal Quality, Usability, Semantic Similarity, Contributor Behavior, Uncertainty, Correctness and Lineage. These quality measures are adopted by the majority of research studies. Also, 17 other important quality measures were obtained: Positional Uncertainty, Currentness, Shape Similarity, Granularity, Compliance, Richness, Ambiguity, Plausibility, Precision, Confidence, Volunteer Credibility, Spatial Error, Time Transition, User Interaction, Objects, Users and Tags. Those quality measures are developed by several reviewed studies that aimed to create new methods for VGI quality assessment. Figure 7 shows the frequency of use of each quality measure.

Hence, we can classify those selected VGI quality measures into three main classes. The first is the heavily used measures that concerns VGI quality measures whose frequency of use reaches or exceeds 10 times. Those measures are used by most research work such as, Positional accuracy, thematic accuracy, completeness, logical consistency and reputation. Previously mentioned criteria show their efficiency in VGI quality assessment. The second, moderately used measures, comprises measures used from 3 TO 8 times. Those criteria are used by some research works separately or combined with others VGI quality indicators. We can quote for example: credibility or trustworthiness. And the third, the weakly used measures, that includes measures which is quoted by one or two research. The frequency of use can reflect the efficiency of measure, but it's not always the case.

Indeed, to fully understand different measures and their use, we propose to classify them into three main categories: spatial measurements, temporal measurements, and relevance measurements.

- Spatial measurements: concerns criterion used to measure spatial features. These indicators assess the credibility of spatial aspects consist of geographic location and spatial relationships in VGI data. Table 5 presents the list of spatial quality measurements and their definitions.
- Temporal measurements: comprises temporal criterion used to measure temporal features. These indicators are used to ensure the quality of the time-

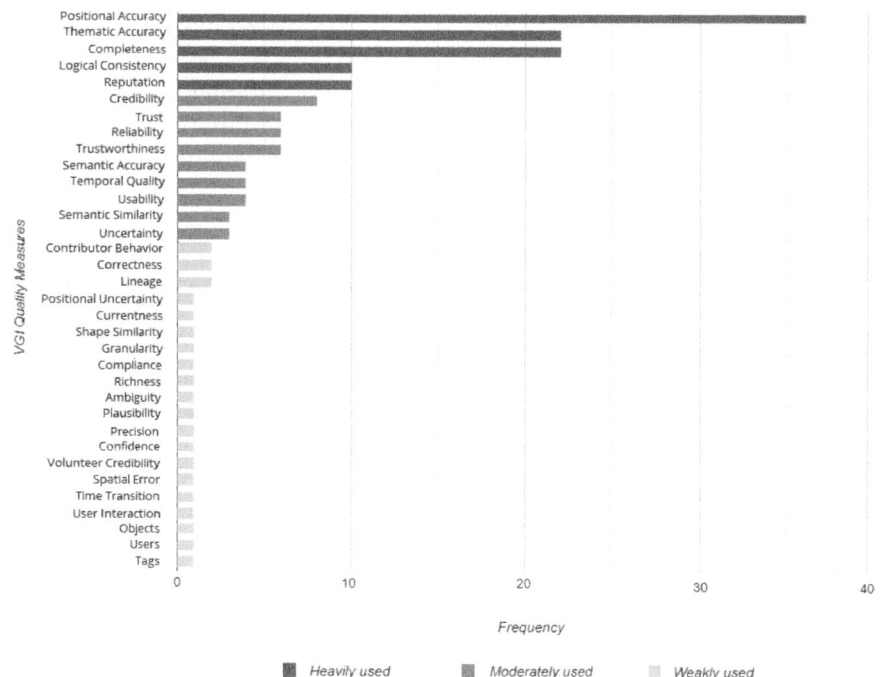

Fig. 7. Frequency of quality measure

related aspects of VGI data. Specifically, it aims to assess data changes over time. Table 6 describes the list of temporal quality measurements.
- Relevance measurements: includes indicators used to assess the pertinence and usefulness of VGI data. It allows the assessment of the importance or significance of features in relation to a particular context or purpose. Relevance measurement offers a subjective evaluation as a complement of spatial and temporal measurement. Table 7 summarizes the list of Relevance quality measurements.

This classification can be useful for future researchers since it allows to identify the strengths and weaknesses of each measure. Categorizing VGI quality measurement allows to analyze and better understand the use of each criterion. Also, that kind of combination was expected in the survey existing measure in order to improve their effectiveness.

4.2 Quality Methods for Assessing VGI

In this section, we review existing quality assessment methods for VGI. We classified these methods according to four analysis axis related to previously identified problematic.

Table 5. Spatial measurements

Spatial measurement	Description
Positional uncertainty	Refers to the distance between the location recorded by an observer for a given taxon and the closest mapped individual of that taxon [36]
Similarity	Quantifies how similar, or dissimilar, concepts are based on their meanings. It computed by looking at the spatial co-occurrence of features [37]
Spatial Error	Refers to the difference between the actual or true location of an object and its estimated or measured location. Spatial error refers to completeness and accuracy together [38]
Spacial Objects	Refers to the number of objects per user: shows the level of users' participation and contribution pattern to determine how many users provided information on how many objects. The higher number of objects per user, the less local knowledge in the dataset [39]
Users	Refers to the number of users in a square kilometer: It explicitly conveys the participation level in the area, where a higher number of users in a square kilometer is an indication of both greater participation and a higher quality dataset [39]
Tags	Refers to the number of tags per object. It shows the level of completeness of map features. The higher the number of tags per object, the higher the level of detail in a map [39]
Shape similarity	Compares the similarity of OSM polygons forms and the corresponding polygons in the OSI reference dataset [40]

Table 6. Temporal measurement

Temporal measurements	Description
Time transition	Analyzes time series of changes provided by contributors used to clarify the geographic distribution of VGI data [41]
Currentness	Evaluates the actuality of the database relative to changes in the real world [25]

4.2.1 Axis 1: VGI Type

The first Axis aims to answer the following question: what is the data source of VGI used by each approach? It concerns the types of VGI to be assessed: collaborative mapping (OpenStreetMap: OSM), participatory sensing, social media. Numerous research works have been proposed to assess VGI quality from various sources.

Table 7. Relevance measurement

Relevance measurements	Description
Reputation	Indicates how much the source is considered reliable within a community [42]
Trustworthiness	Indicates the score that is associated to each version of a geographic feature and that depends on both the sequence of edits that the feature undergoes over time and the reputation of its author [42]
Credibility	Refers to data that exist in the range of known or acceptable values [28]
Reliability	Refers to whether one can trust the data. Reliability is composed of five subdimensions: accuracy, consistency, completeness, integrity, and auditability [28]
Compliance	Indicates the degree of adherence of an attribute, a feature, or a set of features to a given source S, ranging from non-compliance to full compliance [43]
Richness	Refers to the amount and variety of dimensions that are included in the description of the real-world entity [43]
Ambiguity	Refers to notion or object that can have multiple meanings. Ambiguity is due to non-specificity or conflict and arises if there is doubt about how to define or classify an object or phenomenon [44]
Plausibility	Expresses the likelihood of an entity belonging to a specific class [45]
Trust	Refers to data author (author experience and expertise), spatial trust (distance from author's home location to the encoded feature) and temporal trust (the length of time that author spends in any given activity space) [46]
Uncertainty	Indicates the discrepancy between what a database indicates and what actually exists in the real world [47]
Contributor Behavior	Reflects contributors' motivation and individual preferences in selecting mapped features and delineating mapped areas [48]
Correctness	Refers to the degree to which something (features) conforms to a set of specifications, requirements, or expectations [49]
Granularity	Refers to the level of thematic description present in the data, moving from very abstract to very specific concepts [43]
Confidence	Refers to the level of certainty or trustworthiness associated to the feature [50]
Precision	Refers to how close two or more measurements are to each other [51]
Semantic accuracy	Defines as the performance of the system in capturing the meaning of the text and conveying it accurately [52]
Volunteer credibility	Takes observed variables provided by volunteers to compute information on the unobserved (latent) variable [52]
User interaction	Analyzes user interactions through users posts. It analyzes how many users accounts were participating [41]
Lineage	Helps in knowing the source of the data so that any inconsistency is corrected in the source and not in any other instances [19]
Similarity	Quantifies how similar, or dissimilar, concepts are based on their meanings. It is computed by looking at the spatial co-occurrence of features [37]

Table 8. VGI Type

VGI Type	Methods
Collaborative Mapping	[1, 8, 23, 25, 29, 37–43, 45, 46, 48, 49, 51–80]
Participatory Sensing	[1, 10, 11, 29, 43, 46, 49–52, 55, 57, 62, 65, 66, 69, 74, 77, 81–90]
Social Media	[1, 9, 11, 12, 36, 40, 43, 46, 49, 51, 52, 55, 62, 63, 65, 66, 74, 91–93]

The majority of studied searches focuses on the evaluation of VGI collected through OSM wiki[9]. Few methods are reserved to evaluate quality of VGI from social media and participatory sensing. In Table 8, we explored the research works for VGI assessment from numerous sources that use different techniques. Several researchers assess VGI quality from collaborative mapping source. The authors in reference [53] aimed to evaluate the semantic accuracy of the OSM tag also called (features). Its main idea is to develop a methodological framework based on a confusion matrix approach to determine the classification accuracy using all of the information diversity of OSM. Additional research work such as proposed in reference [37] aimed to develop OSM and add new functionalities as a plugin to support quality analysis of VGI. In the same context, new methods and indicators are developed in [54] to evaluate the quality of OSM data. The research work proposes a framework for intrinsic OSM data quality analysis, named OSMAnalyzer. The framework is implemented as a tool and allows anyone to generate information about OSM data quality for a freely selectable area using only OSM's data history.

Research in [61] aims to proposes an approach for improving the semantic quality by using a tag recommended system, called OSMantic, which automatically suggests appropriate tags to contributors during the editing process. Hence, the approach helps reducing the semantic heterogeneity of VGI datasets by assisting contributors to find the most appropriate tags for a given object. Then, implement overall approach into a plugin for the Java OSM editor (JOSM). The authors of reference [67] presented an extension of the Quantum GIS (QGIS) processing toolbox by using existing functionalities and writing new scripts to handle spatial data quality. This allows researchers to assess the completeness of spatial data using intrinsic indicators. The study also proposes a heuristic approach to test the road navigability of OSM data. OSM is also investigated by [69] which uses geometric properties such as the length, distance, orientation, area, shape, buffer overlapped area, complexity, size, and sinuosity to measure data completeness and degree of spatial similarity. Besides OSM, we found numerous research that studies VGI quality collected from social media and other geo-wiki platforms. In [12], a conceptual workflow to collect, clean and process the social media VGI data is proposed before its integration with external authoritative databases for the purpose of risk management. Additionally, authors of refer-

---

[9] https://wiki.openstreetmap.org.

ences [36,91] and [9] proposed solutions for assessing VGI quality from social networks. Other than collaborative mapping and social media sources, there are many research works that assess VGI quality from participatory sensing source. The authors of reference [83] proposes a template for the documentation of VGI from Geobrowsers or Virtual Globes systems namely the Dynamic Metadata for VGI (DM4VGI). The proposed template aims at creating dynamic metadata in order to validate the quality of the VGI, as well as facilitating the interoperability of data collected by several websites with different and dynamic content. Also, it allows users to verify and certify the quality of VGI data, allowing it to be used by any collaborative Geobrowser. Also, reference [81] developed a method for determining the reliability of the VGI about land cover map collected via Geo-Wiki. Similarly, to aid the validation of satellite sensor derived land cover products, the authors of reference [82] explored the potential to estimate the accuracy of VGI collected via Geo-Wiki by analyzing the accuracy of each volunteer or data source. Hence, this work situated at the interface of remote and citizen-sensing. Moreover, there are many studies that offer generic solutions for all types of sources. The authors of reference [62] presented a novel approach for reputation evaluation that is based on the well-known PageRank algorithm for Web pages. Additional research proposed by authors of reference [49] described a conceptual model of indicators (parameters) and practical approaches to automated assess the credibility of VGI including map mashups, Geo Web and crowd-sourced based applications based on machine learning.

4.2.2 Axis 2: VGI Attributes

This axis aims to answer the following question: Which quality attributes can be associated to a selected method? Indeed, there are some methods that use ISO criteria to measure quality while other methods employ additional criteria. In an attempt to summarize the existing VGI quality measures in the literature, we have reviewed the totality of 57 selected paper. As we can see in Table 9 and Table 10, 25% of research implements only ISO quality measures to evaluate VGI quality. While, 75 % of research work try to combine existing criteria to raise the expected VGI quality in various applications. The authors of references [8,54–57,67,68,72,84] and [89] used only ISO quality measures to evaluate VGI quality. By analyzing those quality measures, we noticed that accuracy measurement is frequently used by several authors (60% of studies) to assess OSM quality due to its importance. As shown in Table 9, the authors of reference [63] used all ISO measurement to build their conceptual model. While, the authors of references [82] and [86] choose positional and thematic accuracy measurement to develop their approaches. On the other hand, further research was carried out without using ISO quality measures [39,40,51,61,62,66,85,94] and [74]. They adopted others quality measures to evaluate VGI quality such as credibility, trustworthiness, reputation, semantic similarity, etc. Table 10 summarizes all others quality measurements. Indeed, to measure the trust of VGI, authors in reference [87] proposes a model to assess the quality of spatio-temporal and semantic components and the reliability of the individual producing the information and their collection methods. In reference [65], the authors adopt the notion of trust as an indicator of VGI quality and define the concept of trustworthiness of a VGI record as a function of three main contexts: consistency with habitat, consistency with neighbors, and the reputation of the volunteer. Using fuzzy control system, the quality of an observation is quantified in terms of the level of the trustworthiness of the volunteered species observation. The authors in reference [42] presented a model to measure trustworthiness scores and reputation of contributors by analyzing geometric, qualitative, and semantic aspects during the editing process.

The model was implemented as a Java application using semantic technologies: ontology's have been developed for the representation of OSM data. Using trustworthiness too, the authors of reference [50] introduced a novel hierarchical approach for data quality assessment. The proposed solution focuses on trustworthiness and consists on providing an additional confidence degree for each computed suitability degree. Other quality measures shown their effectiveness in VGI quality assessment trust and trustworthiness. These quality indicators are explored in [91] which proposes a method for automatically creating trustworthiness and reputation scores in order to assess the quality of VGI features. [41] proposes a new approach for the accumulation of OSM data by using OSM Notes and attempts to analyze time transition and user interaction.

Table 9. ISO Quality Measures

ISO Quality Measures	Methods
Positional Accuracy	[1, 8–10, 12, 23, 25, 29, 36–38, 43, 45, 49, 52–54, 56–59, 63, 67, 68, 70–73, 77, 79, 81, 82, 84, 86, 88, 89, 93]
Thematic Accuracy	[1, 9, 10, 12, 25, 29, 43, 45, 49, 52, 53, 63, 67, 68, 70, 81, 86, 88, 89, 93]
Completeness	[8, 10, 12, 23, 25, 29, 38, 43, 48, 49, 54, 56–59, 63, 64, 67–70, 72, 78, 82]
Logical Consistency	[12, 43, 49, 54, 55, 58, 63, 72, 88, 90]
Temporal Quality	[10, 12, 25, 63, 90, 93]
Usability	[25, 63, 88, 92]

As we can deduct from Table 9 and Table 10, many researchers combined ISO quality measures and others Quality measures in their assessment. The authors of reference [25] used positional accuracy, thematic accuracy, semantic accuracy, temporal quality, completeness and usability combined with currentness and lineage to assess VGI quality. More recent work of [43] outlines a multi-faceted framework that includes ISO quality dimensions combined with granularity, compliance and richness. Other studies in reference [58] suggest to assess OSM data using the case study of Heidelberg (Germany) by measuring the degree of completeness, positional accuracy, and semantic accuracy of the dataset were analyzed. Also, reference [71] controlled positional accuracy indicator to decreases the spatial heterogeneity of point features in VGI systems. Reference [38] proposes a semi-automatic approach for quantifying the completeness and accuracy of crowdsourced point data. Other contributions in reference [29] aim to develop a framework for assessing the quality of VGI. An analysis of the last Geo-Wiki tool to be developed is then analyzed with respect to the proposed framework. For this purpose, the authors exploit many quality indicators such as reliability, credibility, positional accuracy, thematic accuracy and completeness. Then, a novel Trust and Reputation Modelling methodology is proposed in [52] to establish VGI data quality. In this work, four quality indicators are examined: thematic accuracy, semantic accuracy, volunteer reputation and positional accuracy. Reference [46] presents 'VGTrust' model to assesses information about a data author, and the spatial and temporal trust associated with the data they create, to produce an overall VGTrust rating metric. This research allows the integration of VGI and authoritative data. Another quality is also frequently used in the analysis of VGI quality adopted by authors in reference [48] and reference [23] known as contributor's behavior. Reference [48] studied contributors' behavior when mapping, such as preferences for objects to map, to understand the characteristics and quality of the data produced. As far as [23] exploits historical map features and contributor behavior to assess VGI quality. Using unsupervised machine learning the ontology of data quality measures, they identify a group of experienced OSM contributors that could be regarded as "power mappers/validators". Reference [73] demonstrates an approach for computing the reliability indicators as tools for assessing OSM data quality using the history of data. The reliability indicator is calculated through criteria such as the number of versions, the number of user participation, temporal variations, and the number of tags editing.

4.2.3 Axis 3: Authoritative Data

Here we aim to answer the following question: what methods can be used to assess the quality of VGI when authoritative data is not available? This allows to know the effectiveness of each method. In Table 11, we have classified methods into two main categories. The first concerns methods that require the presence of reference data to ensure validation. The second concerns methods which does not require reference data.

- Authoritative Data Used
  The majority of research works appealed for reference data during data assessment process. Here, we distinguish methods that not indicate the source of reference data used to validate the proposed approach such as studies proposed in references [1, 10–12, 42, 43, 53, 57, 62, 63, 65, 66, 71, 73, 85, 87, 89, 91] and [74]. However, some others studies indicate the source of authoritative data used. The authors of reference [25] validates their model by comparing different components of French OSM data to a reference institutional database of the French National Mapping Agency (Institut Geographique National, IGN). In the same purpose, the study proposed by [40] was tested by comparing OSM data with ground-truth data for Ireland lakes. Also, the authors

Table 10. Others VGI Quality Measures

Others Quality measures	Methods
Reputation	[10–12, 42, 52, 62, 66, 76, 83, 91, 94]
Credibility	[9, 29, 49, 51, 60, 75, 92–94]
Trust	[1, 46, 49, 52, 87, 88]
Reliability	[1, 10, 29, 73, 81, 87]
Trustworthiness	[12, 42, 50, 65, 91]
Semantic Accuracy	[25, 52, 58, 59, 70]
Semantic Similarity	[37, 53, 61, 76]
Uncertainty	[74, 85]
Contributor Behavior	[23, 48]
Correctness	[49, 64]
Lineage	[25, 70]
Positional Uncertainty	[36]
Currentness	[25]
Shape Similarity	[40]
Granularity	[43]
Compliance	[43]
Richness	[43]
Ambiguity	[45]
Plausibility	[45]
Precision	[51]
Confidence	[50]
Volunteer Credibility	[52]
Spatial Error	[38]
Time Transition	[41]
User Interaction	[41]
Object	[39]
Users	[39]
Tags	[39]

of reference [94] used authoritative data from official, authoritative spatial data infrastructures (SDI) to validate proposed approach. Another approach was evaluated by applying a real-world dataset in the U.S was presented by authors of reference [55]. The authors of reference [56] compared the OSM (as an example of a VGI dataset) with the Integrated Transport Network (ITN) layer of MasterMap from Ordnance Survey (OS), the UK's mapping agency (as an example of a Reference dataset). To demonstrated its effectiveness, the authors of reference [56] suggests a 'feature based' automated method

Table 11. Authoritative Data

Methods	Auth.Data Used	No Auth.Data Used	Types of Authoritative Data
[25]	*		Reference institutional database of IGN
[40]	*		Ground-truth data for Ireland lakes
[94]	*		Authoritative spatial data infrastructures (SDI)
[1]	*		Not indicated
[55]	*		Real world dataset of plant phenology
[56]	*		ITN layer of MasterMap from OS, the UK's mapping agency
[48]	*		A sample area from OSM
[57]	*		Not indicated
[81]	*		GLC-2000, MODIS and Globcover data land cover
[82]	*	*	High quality ground reference data set generated by three experts
[38]	*		National geospatial dataset and ORNL
[83]		*	
[37]		*	
[84]		*	
[58]	*		Administrative data of Germany provided by the FACG
[45]		*	
[51]		*	
[85]	*		Not indicated
[49]		*	
[54]		*	
[59]	*		ATKIS data provided in 2010 by the city of Munich
[8]	*		Official data of Tehran, Iran
[43]	*		Not indicated
[86]		*	
[60]	*		ALKIS and HAZUS catalogues developed by FEMA of the United States
[29]	*		VGI data collection by National Mapping Authorities (NMAs)
[61]		*	
[62]	*		Not indicated
[63]	*		Not indicated
[53]	*		Not indicated
[91]	*		Not indicated
[9]	*		VGI datasets from the Ushahidi projects in Brisbane
[87]	*		Not indicated
[64]	*		OSM reference data
[10]	*		Not indicated
[65]	*		Not indicated
[66]	*		Not indicated
[67]		*	
[68]	*		Data of the National mapping agency in Pakistan

continued

Table 11. continued

Methods	Auth.Data Used	No Auth.Data Used	Types of Authoritative Data
[88]	*		E-Flora record database
[50]		*	
[69]	*		Road networks from the Iran National Cartographic Center (NCC) in 2011
[92]	*		Data from state spatial data infrastructure and irrigation department
[93]		*	
[52]		*	
[70]	*		National road networks from Digital Mapping Technologies, Canada, 2015
[42]	*		Not indicated
[71]	*		Not indicated
[46]	*		Real-world data by New Zealand's mapping organization
[11]	*		Not indicated
[23]	*		DMTI Spatial Inc.™ CanMap® Suite, 2017
[36]	*		INaturalist platform
[12]	*		Not indicated
[41]		*	
[39]		*	
[72]	*		Western Pennsylvania Regional Data Center (WPRDC, data.wprdc.org).
[73]	*		Not indicated
[89]	*		Not indicated
[74]	*		Not indicated
[75]	*		The Ushahidi project: flood which was hit in Brisbane, Australia in 2013
[90]		*	
[76]		*	
[77]		*	
[78]	*		Data provided by three experts from the LOKI project
[79]	*		OSM reference data
[80]		*	

of matching linear data between VGI and a reference dataset to evaluate data completeness of OSM in several case studies in the UK. [48] tested their approach using a sample area from OSM. Additionally, reference [81] compared volunteered land cover information with the GLC-2000, MODIS and Globcover data and to calculate geographically weighted measures of correspondence with VGI data. Furthermore, the authors of reference [82] used a high-quality ground reference data set generated by three experts to evaluate the accuracy of volunteered data. Using a national geospatial dataset as the reference benchmark, the authors of reference [38] tested their approach. The reference data is based upon information from the Department of Education's lists of public and private schools and the Federal government, Oak Ridge National Laboratory (ORNL). The authors of reference [58] evaluated the volunteers' contributions to the OSM project based on comparative investigations with administrative data of Germany provided by the Federal Agency for Cartography and Geodesy. Also, the authors of reference [8] tried to assess the quality of linear VGI features in Tehran, Iran, by comparing them with official data. Reference [60] used ALKIS and HAZUS reference catalogue developed by the Federal Emergency Management Agency of the United States Department of Homeland Security for multi hazard loss estimation in their validation. Then, reference [29] used data from National Mapping Authorities (NMAs). The authors of reference [9] collected reference datasets from the Ushahidi projects in Brisbane. Reference [64] used OSM to extract reference date. The authors of reference [68] used the city of Islamabad in Pakistan as a case study and take the classic approach of comparing VGI data of educational facilities in Pakistan to data collected by the National mapping agency in Pakistan (Survey of Pakistan (SoP)). Reference [88] used 518 authoritative occurrence records extracted from the E-Flora records' database. E-Flora BC (E-Flora) is a biogeographic online atlas of the vascular plants in British Columbia, Canada. Reference [69] used authoritative data provided by the Iran National Cartographic Center (NCC) in 2011. Also, the authors of reference [92] adopted authoritative data collected from the state spatial data infrastructure and irrigation department. Moreover, reference [70] made a comparison of the overall OSM quality in Canada with Digital Mapping Technologies Inc. (DMTI). While, reference [46] used a real-world data by New Zealand's mapping organization. Reference [23] used dataset from DMTI Spatial Inc, 2017 as an authoritative reference due to its availability and popularity in the region. In the same context, reference [36] used authoritative data from i-Naturalist platform. And finally, reference [72] used reference data from Western Pennsylvania Regional Data Center (WPRDC, data.wprdc.org).

- No Authoritative Data Used
  A few numbers of validates research are mentioned without using reference data. the authors of references [49,51,54,61,86,93] and [52] proposed approaches to adopt when reference data is unavailable. Other authors of references [37,41,45,50,67,83] and [84] made the evaluation of VGI based on comparing the existing data of the same place with each other according to the metadata that their creators have given. They used old and recent OSM data to made the comparison.

4.2.4 Axis 4: Application Domains

This axis aims to answer the following question: what are the domains of application of each approach? This allows to know if the method is approved and tested in real life situation and then the application area of each one. The study of VGI quality is important for several reasons. It informs decision-making, promote citizen engagement, and support the integration of data from various sources.

In the last years, authors seek to build robust approaches to deal with the issue of VGI quality by analyzing various case study and extending the application areas. After analyzing reviewed papers, we notice that all selected studies were developed in principally eleven domains: Mapping, Disaster management, Citizen science, Social analytic, Sport, Tourism, Educational planning, Land use, Agriculture and Roads networks. Some other studies were classified as Generic since they can be applied to several areas. Table 12 summarizes all the domains invoked in reviewed papers.

- Generic
  Some authors proposed a generic approach without any domain specification rather than a specific or specialized approach. These general methods can be useful in a variety of contexts, situations or problems as it allows for flexibility and adaptability. The implementation of the framework proposed by authors of references [41, 42, 49, 74] can be extended in many directions.
- Mapping
  From the table, we noted that Mapping is the largest domain, comprising 12 studies. Several researches proposed to maintain VGI quality during mapping action in OSM. Accordingly, the authors of reference [58] proposed an approach in OSM mapping that categories OSM contributors (mappers) based on the quantity and quality of their shared data. As such "beginners", "regular mappers", "intermediate mappers", "experts", and "professional mappers" are identified. Additionally, the authors of reference [40] presented an approach that aims on mapping eleven countries and regions using OSM. In the same context, the authors of references [8, 37, 39, 43, 45, 53, 57, 73, 84, 89] and [1] proposed a VGI quality assessments approaches applied in various mapping activities.
- Disaster Management
  Nowadays, VGI is a useful tool in the case of crisis management and emergency response. In fact, decision making is the main goal of the analysis of VGI quality specifically in crisis situation. As shown in Table 12, a large number of research deals with disaster management domain [12, 51, 64] and [46]. Disaster management involves a range of activities and measures aimed at reducing the risks of disasters, including hazard assessment, risk analysis, emergency planning, and the development of early warning systems. It also includes the provision of emergency services and support to affected communities during and after disasters, such as search and rescue, medical assistance, shelter, and basic needs like food and water. In this context, the authors of reference [94] presented a complete conceptual workflow for the assessment of VGI credibility for the use in a crisis management context. Furthermore, authors of reference [85] assessed the uncertainties of VGI submitted by users in the emergency response phase of disaster situations. The case study takes the example of data created in response to the 2010 Haiti earthquake distributed via the Ushahidi web platform geospatial database. Also, the authors of reference [60] presented first results towards a method for assessing the quality of OSM for the specific in order to identifying assets of critical infrastruc-

ture in support of emergency planning. Reference [63] outlines a conceptual model to assess the quality of volunteered geographic information for the tasks related to flood management. For the same purpose, the authors of reference [93] assesses the credibility of the Australian Broadcasting Corporation's Ushahidi CrowdMap dataset. The approach was tested using text-based content submitted by volunteers during the flood event. Then, Reference [9] aims to develop a method to calculate the credibility scores of VGI for disaster response. By collecting datasets from two extreme flood events in 2011 and 2013 from Brisbane (Australia), the model shows a great potential to be used by emergency management sectors for efficient and rapid response, decision-making, and coordination. The objective of the study suggested by reference [61] is the quality analysis of remote mapping and the evaluation of the correctness and completeness of OSM data via a comparison of data contributions by the students versus OSM reference data for the purpose of disaster management. Similarly, the authors of reference [92] proposed a statistical framework developed to validate the credibility of citizen generated spatial data for emergency decision making.

- Citizen Science
  Citizen science is a domain of scientific research that involves collaboration between professional scientists and volunteers from the general public. It represents an innovative and inclusive domain to scientific research that has the potential to advance knowledge. In this context, the authors of reference [29, 86] and [38] proposed approaches to deal with the topic of citizen science. furthermore, the authors of reference [88] described a fuzzy model for intrinsic quality assessment of VGI on species occurrences obtained by Citizen Science (CS) biodiversity monitoring programs. Also, [10] outlined a multidimensional framework for the assessment of conceptual quality to facilitate the communication over the semantic gulf that separates producers and consumers. Reference [65] described a conceptual model for quality assurance of species occurrence observations in citizen science projects.
- Social Analysis
  Social analysis is a broad domain that encompasses the study of social phenomena, including human behavior, social structures, institutions, and culture. It involves the systematic examination and interpretation of social interactions and patterns, with the aim of understanding social relations and making informed decisions about social issues. In this direction, the authors of reference [87] analyzed through a case study with the collection of VGI representing walking and biking tracks. Additionally, the authors of references [83, 91] and [71] developed a VGI system that aims at collecting landscape description information to Know how people see and perceive a landscape imparts additional information about different characteristics of a landscape.
- Tourism
  Further research concerns the domain of tourism planning and e-tourism. The authors of reference [66] developed an approach for VGI quality evaluation approved in the context of geospatial decision support systems and in recom-

mending tourist itineraries. Also, the authors of reference [11] tested his proposed model via an applicable scenario for recommending tourist itineraries.
- Educational Planning
  Another domain of application explored by authors of reference [68] which concerns educational planning. In this study, authors examine the quality of OSM data for the purpose of educational planning, they investigates the city of Islamabad in Pakistan and take the classic approach of comparing the spatial, semantic and thematic data of educational facilities collected in OSM to data collected by the National mapping agency in Pakistan as reference data.
- Land Use
  Land use refers to the way in which land is utilized for various purposes, such as residential, commercial, agricultural, industrial, and recreational activities. It is concerned with the study of land use patterns, the impact of land use on the environment and human societies, and the planning and management of land use. In this context, The authors of references [82] and [52] exploits VGI quality assessment in Land-use and Land Administration Systems. Also, reference [81] shown that VGI about land cover can be used in formal analyses when it is linked to control data (locations where the land cover was known.
- Agriculture
  Exploring other domains handled by selected paper, we find that VGI quality is also applied in the agriculture areas. The approach of reference [55] evaluated by applying it to a real-world dataset, with observations of plant phenology, made by non-expert volunteers in the U.S.A. As well as the approach proposed by authors of reference [36] that studying positional uncertainty and positional accuracy to evaluate VGI quality in iNaturalist for vegetation mapping.
- Road Networks
  Road network analysis involves the study of the spatial arrangement of roads and the patterns of traffic flow. This analysis can be used to identify areas of congestion, safety concerns, and opportunities for improvement. In this context, the authors of references [25, 48, 50, 67, 69, 70] and [23] studied rod networks to presents the effectiveness of their approaches. Indeed, Reference [70] aims to evaluate the extrinsic quality of the Canadian OSM street networks in terms of positional accuracy. The authors of reference [72] proposed a comprehensive quality assessment framework for VGI linear features and examines the spatial auto-correlation and semantic correlation. The OSM road network (sidewalks, buildings, points of interest) of Allegheny County, Pennsylvania (USA) was selected as an example to test the proposed framework. Also, the authors of reference [56] proposed a method applied in urban and rural areas of Greater London and west of Newcastle respectively to see how effective Road networks of selected area.

Table 12. Application Domains

Domains	Methods
Generic	[41, 42, 49, 74, 76]
Mapping	[1, 8, 37, 39, 40, 43, 45, 53, 57, 58, 73, 84, 89]
Disaster Management	[9, 12, 46, 51, 60, 61, 63, 64, 75, 85, 92–94]
Citizen Science	[10, 29, 38, 59, 65, 78–80, 86, 88, 90]
Social Analysis	[66, 71, 83, 87, 91]
Tourism	[11]
Educational Planning	[68]
Land Use	[52, 77, 81, 82]
Agriculture	[36, 55]
Road Networks	[25], [23, 48, 50, 54, 56, 67, 69, 70, 72]

5 Discussion and Future Perspectives

Evaluate the quality of volunteered data and recover the knowledge hidden in the information is today a great challenge. In this paper, we have provided a recent overview of quality assessment studies for VGI. In order to understand these studies, we determined our specific scope by formulating four different research questions, (i) What are the data source of VGI used by each approach? (ii) Which quality attributes can be associated to a selected method? (iii) What methods can be used to assess the quality of VGI when authoritative data is not available? (iv) What are the domains of application of each approach? This review studied research papers published between 2010 and 2021. Contrary to existing studies, this work focuses on recent methods used to assess the quality of VGI and quality attributes adopted by each method take into account the related sources of VGI. Indeed, for each reviewed paper, we list a set of quality measures and we classified them in accordance with the type of VGI (i.e., social media, Participatory sensing, and collaborative mapping).

5.1 Sources of VGI

In this survey, we adopted a taxonomy that took into account the types of VGI source (Collaborative mapping, Social media, Participatory sensing). After analyzing the selected papers in details, we identified that the majority of reviewed papers are intended principally for VGI out coming from collaborative mapping such as OSM (around 65 % of papers). The rest of papers includes VGI quality assessment methods produced in social media and participatory sensing tools. We also noticed that there exist few research works that include methods for assessing the quality of all types of VGI (around 20 % of papers). The authors of references [1, 55] and [49] presents a work that includes solution for quality assessment of VGI collected from all types of sources.

5.2 Assessment Criteria

Various metrics and indicators are identified in this work. We can categorize them in two main class. the first-class concerns quality elements defined by ISO/TC 2117 as an international standard that defines the measures of geographic data quality (completeness, positional accuracy, thematic accuracy, logical consistency, temporal quality and usability). And the second-class concerns other quality elements proposed by researchers to evaluate VGI quality. However, some studies used all ISO quality metric to assess VGI quality such as reference [63]. However, some other studies choose to combine only two or three of them in their evaluation and ignore the rest such as references [57] and [86] that adopt only completeness and accuracy for their quality assessment. On the other hand, we noticed that some studies use a combination of ISO quality elements to create new quality measurement such as reference [38] that create a new quality measure namely spatial error. To calculate the score of spatial error, the author uses completeness and accuracy. Furthermore, the authors of reference [52] used credibility to calculate trust and reputation measures. In fact, the concept of VGI quality is strongly related to the term of "credibility". Data credibility is also related to several quality measurements, including trust, reliability, accuracy, reputation, authority and competence [95]. The authors of reference [60] employed credibility to describe VGI quality. Despite the number of measures introduced and applied, according to the studies analyzed, there is no study that assesses all the elements of VGI quality. A generic approach that will produce integrated results is missing.

5.3 Application Domains

VGI can be obtained from several sources and deal with numerous issues in real life mainly in crisis management [9,63,85,92,94] and [12], Ecological problems [36], developing countries and smart cities [52], Health Care [10], Social Reporting [96], etc. In this review, we detected that most of research works focus on the application domain of disaster management and emergency response. A few of reviewed papers (7 %) propose VGI quality assessment method that can be employed in any field of study [41,42,49,74]. Mapping activities is also an important domain of application for VGI quality assessment [1,40] and [57].

5.4 Assessment Methods and Tools

While most methods have been conducted to deal with the issues of quality in stage of evaluation, fewer methods tackle the issue of data quality in the different stages of VGI data creation, collection and evaluation. In this context, [96] propose a tool that implement PageRank algorithms to evaluate quality on social reporting scenarios, in which citizens submit geotagged reports about observations of real-world-events. This work concentrates on data quality in the different stages of VGI data creation, collection and evaluation. There are many other works that propose tools which aids to assess VGI quality. The authors of reference [14] explore a list of tools that aims to develop VGI system and that can be used in emergency situations, when the data must be collected and distributed almost in real time. It compared VGI data with data from governmental agencies and concluded the VGI data quality such as Ushahidi Platform and ClickOnMap Platform. In parallel with the tools cited above, there are also a number of tools that have been developed to evaluate VGI quality. These tools are generally dedicated to help OSM users to assess the quality and the reliability of information provided by their community. These tools allow detecting errors in the OSM data and identifying erroneous modification. We can classify these tools into two main categories. The first includes error detection that identify potential errors and inaccuracies in the OSM data. Moreover, the second includes monitoring tools that help to identify changes and erroneous modification.

5.5 Research Directions

Several research perspectives can be mentioned regarding the conclusions of this study:

- Although the great number of proposed studies, however, the literature review reveals that none of the existing approach involves in the same time the assessment of source, volunteer and content of VGI data. A research study that aims to highlight those three actors of VGI production process will be important to be explored in the future. That can help to support researchers to have a global view of the literature. Indeed, in the evaluation process, we should be aware of the heterogeneity of VGI and the limitations of current techniques to ensure quality outcomes. Unlike in existing studies, we describe how each method works and what it's the quality attributes related to each method. However, in contrast with existing works, this taxonomy summarizes what methods can be employed to assess the quality of VGI when authoritative data is not available. Then, we noted that 43 studies (73%) used authoritative data in their validation phase, while 16 studies (27%) validate their approaches without any authoritative data. Only one study can be validated with or without any authoritative data. By doing this, we aim to highlight the effectiveness and the limitation of each method. And then, conduct researchers to identify shortcomings to improve them or to design their own solutions.
- As we can deduct from Fig. 7, there are many quality attributes which has a very low frequency of use (Frequency of use equal to one). Each quality attribute adopted only by one of the selected papers. However, those attributes show their effectiveness and efficiency. Existing approaches required considerable configuration efforts to make them suitable for additional data quality criteria integration. As presented previously, there have been a large number of studies and empirical research (70%) adopted ISO quality in their assessment. However, there are some general limitations of ISO quality measures when applied to VGI data. The first limit concerns the applicability of ISO quality measure to VGI data. ISO quality measures are designed primarily for use in industries and contexts that involve formal organizations and

structured processes. The nature of VGI data collection is often more informal and decentralized, which can make it difficult to apply ISO quality measures in a meaningful way. The second limit concern the scalability and the subjectivity of VGI data. VGI data is often subjective and context-dependent, which can make it difficult to establish objective quality measures. Also, VGI data is often collected at a much larger scale than traditional data sources, which can make it difficult to apply ISO quality measures in a practical way. The volume and velocity of VGI data can overwhelm traditional validation processes, requiring new and innovative approaches to quality assurance. Therefore, as a recommendation for future research in this topic, we suggest to develop a solid framework for automatically assist users in their evaluation process that adopt new suitable and flexible indicators.
- Other limitations is mainly related to the rapid growth of the available online sources and how to cope with the performance of the provided solutions. Thereafter, in this paper, we classified methods with regard to the type of collaborative activity. From our analysis, we noted that Collaborative mapping is the largest type of VGI used, comprising 39 studies (67%). The second type is participatory sensing, comprising 26 studies (45%). The fewer type is social media, comprising 18 studies (31%). And finally, we noted that just 9 studies (15%) focused on all types of VGI. Future works must focus on new tools for generating volunteered geographic information such as social networks tools and trying to develop adequate quality indicators given the importance of this information source and the lack of work in this field. Additionally, we classified methods in order to the application domain. We noted that some of the most common application domains include mapping with 13 studies (22%), disaster management with 12 studies (20%) and road networks with 10 studies (17%). Other domains are also invoked such as social analysis, citizen science, agriculture and tourism. Existing approaches may require some update to keep up with the broad range of application domains. Hence, there is an essential need for developing a solid framework for automatically assist users in their evaluation process that take into account the diversity of existing areas. Overall, this work presents a basis on which we can build our future work.

5.6 Future Perspectives

This paper provides a comprehensive summary and synthesis of existing literature on a VGI data quality assessment. It consolidates and organize information of various studies, helping researchers understand the current state of knowledge in the field of VGI and data quality. In fact, our paper identifies gaps in current research to help researchers contributing to the development of their studies and guide them toward novel and relevant works. Indeed, it offers an easy-to-interpret database to grasp the essentials without having to go through numerous individual studies. Moreover, the discussion provided in our review can be beneficial for researchers seeking guidance and inspire them by presenting new ideas, perspectives, and unanswered questions. As part of future efforts, it is intended to develop a framework for quality evaluation of VGI data collected from social media. This framework must be able to analyses the source of information, contributors, volunteers and content Thus, to ensure this assessment, we aim to develop a set of specific criteria together with a combination of already existing quality dimension. Likewise, our future work will illustrate how our system can be used to assess the quality of VGI in social media in order to provide relevant crisis management, assistance for decision making and planning assistance interventions in case of disaster.

6 Conclusion

The concept of crowdsourcing in general and VGI in particular constitute a great challenge as regards the verification and the validation of information quality provided voluntarily by users. Data quality assessment is a crucial step to determine their relevance and their representativeness. In fact, the use of erroneous data, in any context, can cause harmful consequences. To overcome these problems, many works propose solutions and mechanisms to ensure the quality of these data. Initial research in this context involves voluntary information as a whole. They are generally aimed at assessing the quality of data generated voluntarily on the web based on specific performance criteria. More developed studies concern geographic information. Indeed, [ISO 1994][10] defines the assessment or quality control of a geographic database as "measures, examinations, tests, calibration of one or more characteristics of a product and comparisons with the specified requirements in order to establish their conformity". In this paper, we have presented, to the best of our knowledge, the main existing works for VGI quality assessment methodologies and tools. The goal of this survey is to obtain a clear understanding overview of existing approaches, in particular in terms of quality criteria and type of volunteered data. We proceeded with an analysis to highlight the major capacities and shortcomings of these approaches. Our study was discussed according the main data quality criteria. We have also classified these methods according to VGI types. The first concerns VGI produced by collaborative mapping tools, the second concerns VGI produced in social media and the third concerns VGI produced by participatory sensing. We have chosen to review studies published between 2010 and 2021. This work constitutes a base for addressing limitations and a first step for improving the existing methods and paves for new contributions on this topic.

References

1. Goodchild, M.F., Li, L.: Assuring the quality of volunteered geographic information. Spat. Stat. **1**, 110–120 (2012)
2. Fonte, C.C., et al.: Assessing VGI data quality. Mapp. Citizen Sens. 137–163 (2017)
3. Haworth, B., Bruce, E.: A review of volunteered geographic information for disaster management. Geogr. Compass **9**, 237–250 (2015)

---

[10] ISO 8402:1994 https://www.iso.org/standard/20115.html.

4. Kaewkitipong, L., Chen, C., Ractham, P.: Lessons learned from the use of social media in combating a crisis: a case study of 2011 Thailand flooding disaster (2012)
5. Chatfield, A.T., Brajawidagda, U.: Twitter early tsunami warning system: a case study in Indonesia's natural disaster management, pp. 2050–2060. IEEE (2013)
6. Shah, A.A., Ravana, S.D., Hamid, S., Ismail, M.A.: Web credibility assessment: affecting factors and assessment techniques (2015)
7. Antoniou, V., Skopeliti, A.: Measures and indicators of VGI quality: an overview. ISPRS Ann. Photogram. Remote Sens. Spat. Inf. Sci. **2** (2015)
8. Eshghi, M., Alesheikh, A.: Assessment of completeness and positional accuracy of linear features in volunteered geographic information (VGI). Int. Arch. Photogram. Remote Sens. Spat. Inf. Sci. **40**, 169 (2015)
9. Hung, K.-C., Kalantari, M., Rajabifard, A.: Methods for assessing the credibility of volunteered geographic information in flood response: a case study in Brisbane, Australia. Appl. Geogr. **68**, 37–47 (2016)
10. Langley, S.A., Messina, J.P., Moore, N.: Using meta-quality to assess the utility of volunteered geographic information for science. Int. J. Health Geogr. **16**, 1–11 (2017)
11. Jabeur, N., Karam, R., Melchiori, M., Renso, C.: A comprehensive reputation assessment framework for volunteered geographic information in crowdsensing applications. Pers. Ubiquit. Comput. **23**, 669–685 (2019)
12. El Hatimi, B., Oulidi, H.J., Fadil, A.: Quality assessment in volunteered geographic information for risk management applications, pp. 1–4. IEEE (2020)
13. Senaratne, H., Mobasheri, A., Ali, A.L., Capineri, C., Haklay, M.: A review of volunteered geographic information quality assessment methods. Int. J. Geogr. Inf. Sci. **31**, 139–167 (2017)
14. Câmara, J.H.S., Lisboa-Filho, J., de Souza, W.D., Pereira, R.O.: Quality attributes and methods for VGI. In: Gervasi, O., et al. (eds.) ICCSA 2016. LNCS, vol. 9788, pp. 306–321. Springer, Cham (2016). https://doi.org/10.1007/978-3-319-42111-7_24
15. Degrossi, L.C., Porto de Albuquerque, J., dos Santos Rocha, R., Zipf, A.: A framework of quality assessment methods for crowdsourced geographic information: a systematic literature review (2017)
16. Degrossi, L.C., Porto de Albuquerque, J., Santos Rocha, R.D., Zipf, A.: A taxonomy of quality assessment methods for volunteered and crowdsourced geographic information. Trans. GIS **22**, 542–560 (2018)
17. Medeiros, G., Holanda, M.: Solutions for data quality in GIS and VGI: a systematic literature review. New Knowl. Inf. Syst. Technol. **1**, 645–654 (2019)
18. Bordogna, G., Carrara, P., Criscuolo, L., Pepe, M., Rampini, A.: On predicting and improving the quality of volunteer geographic information projects. Int. J. Digit. Earth **9**, 134–155 (2016)
19. Ramasamy, A., Chowdhury, S.: Big data quality dimensions: a systematic literature review. JISTEM-J. Inf. Syst. Technol. Manage. **17** (2020)
20. Ardagna, D., Cappiello, C., Samá, W., Vitali, M.: Context-aware data quality assessment for big data. Futur. Gener. Comput. Syst. **89**, 548–562 (2018)
21. Salvatore, C., Biffignandi, S., Bianchi, A.: Social media and twitter data quality for new social indicators. Soc. Indic. Res. **156**, 601–630 (2021)
22. Albuquerque, J.P.D., Fonte, C., Almeida, J.-P.D., Cardoso, A.: How volunteered geographic information can be integrated into emergency management practice? First lessons learned from an urban fire simulation in the city of Coimbra, 269-276 (2016)

23. Jacobs, K.T., Mitchell, S.W.: OpenStreetMap quality assessment using unsupervised machine learning methods. Trans. GIS **24**, 1280–1298 (2020)
24. Dama. Defining data quality dimensions (2013)
25. Girres, J.-F., Touya, G.: Quality assessment of the French OpenStreetMap dataset. Trans. GIS **14**, 435–459 (2010)
26. Zielstra, D., Hochmair, H.H., Neis, P.: Assessing the effect of data imports on the completeness of OpenStreetMap-a United States case study. Trans. GIS **17**, 315–334 (2013)
27. Mas, J.-F., et al.: A suite of tools for assessing thematic map accuracy. Geogr. J. **2014** (2014)
28. Cai, L., Zhu, Y.: The challenges of data quality and data quality assessment in the big data era. Data Sci. J. **14** (2015)
29. Fonte, C., et al.: VGI quality control. ISPRS Ann. Photogram. Remote Sens. Spat. Inf. Sci. **2**, 317–324 (2015)
30. ISO. ISO 19157: 2013 geographic information - data quality (2013)
31. Mohammadi, N., Malek, M.: Artificial intelligence-based solution to estimate the spatial accuracy of volunteered geographic data. J. Spat. Sci. **60**, 119–135 (2015)
32. Ali, A.L., Schmid, F.: Data quality assurance for volunteered geographic information. In: Duckham, M., Pebesma, E., Stewart, K., Frank, A.U. (eds.) GIScience 2014. LNCS, vol. 8728, pp. 126–141. Springer, Cham (2014). https://doi.org/10.1007/978-3-319-11593-1_9
33. Bishr, M., Kuhn, W.: Trust and reputation models for quality assessment of human sensor observations. In: Tenbrink, T., Stell, J., Galton, A., Wood, Z. (eds.) COSIT 2013. LNCS, vol. 8116, pp. 53–73. Springer, Cham (2013). https://doi.org/10.1007/978-3-319-01790-7_4
34. Kesler, C., De Groot, R.T.A.: Trust as a proxy measure for the quality of volunteered geographic information in the case of OpenStreetMap. Geogr. Inf. Sci. Heart Eur. 21–37 (2013)
35. Bodnar, T., Tucker, C., Hopkinson, K., Bilén, S.G.: Increasing the veracity of event detection on social media networks through user trust modeling, pp. 636–643. IEEE (2014)
36. Uyeda, K.A., Stow, D.A., Richart, C.H.: Assessment of volunteered geographic information for vegetation mapping. Environ. Monit. Assess. **192**, 554 (2020)
37. Vandecasteele, A., Devillers, R.: Improving volunteered geographic data quality using semantic similarity measurements. Int. Arch. Photogram. Remote Sens. Spat. Inf. Sci. **1**, 143–8 (2013)
38. Jackson, S.P., et al.: Assessing completeness and spatial error of features in volunteered geographic information. ISPRS Int. J. Geo Inf. **2**, 507–530 (2013)
39. Forati, A.M., Ghose, R.: Volunteered geographic information users contributions pattern and its impact on information quality (2020)
40. Mooney, P., Corcoran, P., Winstanley, A.C.: Towards quality metrics for OpenStreetMap, pp. 514–517 (2010)
41. Seto, T., Kanasugi, H., Nishimura, Y.: Quality verification of volunteered geographic information using OSM notes data in a global context. ISPRS Int. J. Geo Inf. **9**, 372 (2020)
42. Fogliaroni, P., D'Antonio, F., Clementini, E.: Data trustworthiness and user reputation as indicators of VGI quality. Geo-Spat. Inf. Sci. **21**, 213–233 (2018)
43. Ballatore, A., Zipf, A.: A conceptual quality framework for volunteered geographic information. In: Fabrikant, S.I., Raubal, M., Bertolotto, M., Davies, C., Freundschuh, S., Bell, S. (eds.) COSIT 2015. LNCS, vol. 9368, pp. 89–107. Springer, Cham (2015). https://doi.org/10.1007/978-3-319-23374-1_5

44. Khalfi, B.: Modélisation et construction des bases de données géographiques floues et maintien de la cohérence de mod'eles pour les sgbd sql et nosql. Université PARIS (2017)
45. Ali, A.L., Schmid, F., Al-Salman, R., Kauppinen, T.: Ambiguity and plausibility: managing classification quality in volunteered geographic information, pp. 143–152 (2014)
46. Severinsen, J., de Roiste, M., Reitsma, F., Hartato, E.: VGTrust: measuring trust for volunteered geographic information. Int. J. Geogr. Inf. Sci. **33**, 1683–1701 (2019)
47. Goodchild, M.F.: Academic pursuits-uncertainty: the Achilles heel of GIS? Geo Info Systems **8**, 50–52 (1998)
48. Bégin, D., Devillers, R., Roche, S.: Assessing volunteered geographic information (vgi) quality based on contributors' mapping behaviours. Int. Arch. Photogramm. Remote Sens. Spat. Inf. Sci **2013**, 149-154 (2013)
49. Idris, N.H., Jackson, M., Ishak, M.: A conceptual model of the automated credibility assessment of the volunteered geographic information, vol. 18, p. 012070. IOP Publishing (2014)
50. De Tré, G., et al.: Data quality assessment in volunteered geographic decision support. In: Bordogna, G., Carrara, P. (eds.) Mobile Information Systems Leveraging Volunteered Geographic Information for Earth Observation. ESDM, vol. 4, pp. 173–192. Springer, Cham (2018). https://doi.org/10.1007/978-3-319-70878-2_9
51. Bimonte, S., Boucelma, O., Machabert, O., Sellami, S.: From volunteered geographic information to volunteered geographic OLAP: a VGI data quality-based approach. In: Murgante, B., et al. (eds.) ICCSA 2014. LNCS, vol. 8582, pp. 69–80. Springer, Cham (2014). https://doi.org/10.1007/978-3-319-09147-1_6
52. Moreri, K.K., Fairbairn, D., James, P.: Volunteered geographic information quality assessment using trust and reputation modelling in land administration systems in developing countries. Int. J. Geogr. Inf. Sci. **32**, 931–959 (2018)
53. Albakri, M.M.: Semantic similarity assessment of volunteered geographic information. J. Eng. **22**, 215–229 (2016)
54. Barron, C., Neis, P., Zipf, A.: A comprehensive framework for intrinsic OpenStreetMap quality analysis. Trans. GIS **18**, 877–895 (2014)
55. Yanenko, O., Schlieder, C.: Enhancing the quality of volunteered geographic information: a constraint-based approach. In: Gensel, J., Josselin, D., Vandenbroucke, D. (eds.) Bridging the Geographic Information Sciences. Lecture Notes in Geoinformation and Cartography, pp. 429–446. Springer, Heidelberg (2012). https://doi.org/10.1007/978-3-642-29063-3_23
56. Koukoletsos, T., Haklay, M., Ellul, C.: Assessing data completeness of VGI through an automated matching procedure for linear data. Trans. GIS **16**, 477–498 (2012)
57. Karimipour, F., Esmaeili, R., Navratil, G.: Cartographic representation of spatial data quality parameters in volunteered geographic information (2013)
58. Arsanjani, J.J., Barron, C., Bakillah, M., Helbich, M.: Assessing the quality of openstreetmap contributors together with their contributions, pp. 14–17 (2013)
59. Fan, H., Zipf, A., Fu, Q., Neis, P.: Quality assessment for building footprints data on OpenStreetMap. Int. J. Geogr. Inf. Sci. **28**, 700–719 (2014)
60. Herfort, B., Eckle, M., de Albuquerque, J.P., Zipf, A.: Towards assessing the quality of volunteered geographic information from OpenStreetMap for identifying critical infrastructures. Citeseer (2015)
61. Vandecasteele, A., Devillers, R.: Improving volunteered geographic information quality using a tag recommender system: the case of OpenStreetMap. OpenStreetMap GISci.: Experiences Res. Appl. 59–80 (2015)

62. Lodigiani, C., Melchiori, M.: A PageRank-based reputation model for VGI data. Procedia Comput. Sci. **98**, 566–571 (2016)
63. de Albuquerque, J.P., Fan, H., Zipf, A.: A conceptual model for quality assessment of VGI for the purpose of flood management, pp. 14–17 (2016)
64. Klonner, C., Eckle, M., Usón, T., Höfle, B.: Quality improvement of remotely volunteered geographic information via country-specific mapping instructions (2017)
65. Vahidi, H., Klinkenberg, B., Yan, W.: A fuzzy system for quality assurance of crowdsourced wildlife observation geodata, pp. 55–58. IEEE (2017)
66. Gusmini, M., Jabeur, N., Karam, R., Melchiori, M., Renso, C.: Evaluating reputation in VGI-enabled applications (2017)
67. Sehra, S.S., Singh, J., Rai, H.S.: Assessing OpenStreetMap data using intrinsic quality indicators: an extension to the QGIS processing toolbox. Future Internet **9**, 15 (2017)
68. Muzaffar, H.M., Tahir, A., Ali, A., Ahmad, M., McArdle, G.: Quality assessment of volunteered geographic information for educational planning, pp. 76–96. IGI Global (2017)
69. Chehreghan, A., Ali Abbaspour, R.: An evaluation of data completeness of VGI through geometric similarity assessment. Int. J. Image Data Fusion **9**, 319–337 (2018)
70. Zhang, H., Malczewski, J.: Accuracy evaluation of the Canadian Openstreetmap road networks. Int. J. Geospat. Environ. Res. **5** (2017)
71. Ibrahim, M.H., Darwish, N.R., Hefny, H.A.: An approach to control the positional accuracy of point features in volunteered geographic information systems. Int. J. Adv. Comput. Sci. Appl. **10** (2019)
72. Wu, H., et al.: A comprehensive quality assessment framework for linear features from volunteered geographic information. Int. J. Geogr. Inf. Sci. **35**, 1826–1847 (2021)
73. Teimoory, N., Ali Abbaspour, R., Chehreghan, A.: Reliability extracted from the history file as an intrinsic indicator for assessing the quality of OpenStreetMap. Earth Sci. Inform. **14**, 1413–1432 (2021)
74. Bordogna, G.: A semantic approach for quality assurance and assessment of volunteered geographic information. Information **12**, 492 (2021)
75. Safariallahkheili, Q., Malek, M.R.: A method for assessing the credibility of volunteered geographic information in case of flood crisis. Procedia Comput. Sci. **207**, 1611–1622 (2022)
76. Zhao, Y., Wei, X., Liu, Y., Liao, Z.: A reputation model of OSM contributor based on semantic similarity of ontology concepts. Appl. Sci. **12**, 11363 (2022)
77. Foody, G., Long, G., Schultz, M., Olteanu-Raimond, A.-M.: Assuring the quality of VGI on land use and land cover: experiences and learnings from the landsense project. Geo-Spat. Inf. Sci. 1–22 (2022)
78. Ullah, T., Lautenbach, S., Herfort, B., Reinmuth, M., Schorlemmer, D.: Assessing completeness of OpenStreetMap building footprints using mapswipe. ISPRS Int. J. Geo Inf. **12**, 143 (2023)
79. Kilic, B., Hacar, M., Gülgen, F.: Effects of reverse geocoding on OpenStreetMap tag quality assessment. Trans. GIS **27**, 1599–1613 (2023)
80. Azariasgari, E., Hosseinali, F.: Evaluating the VGI users' level of expertise: an application of statistical and artificial neural network approaches. Int. J. Appl. Geospat. Res. **14** (2023)
81. Comber, A., et al.: Using control data to determine the reliability of volunteered geographic information about land cover. Int. J. Appl. Earth Obs. Geoinf. **23**, 37–48 (2013)

82. Foody, G.M., et al.: Assessing the accuracy of volunteered geographic information arising from multiple contributors to an internet based collaborative project. Trans. GIS **17**, 847–860 (2013)
83. de Souza, W.D., Lisboa Filho, J., Vidal Filho, J.N., Câmara, J.H.: DM4VGI: a template with dynamic metadata for documenting and validating the quality of volunteered geographic information, pp. 1–12. Citeseer (2013)
84. Esmaili, R., Naseri, F., Esmaili, A.: Quality assessment of volunteered geographic information. Am. J. Geogr. Inf. Syst. **2**, 19–26 (2013)
85. Camponovo, M.E., Freundschuh, S.M.: Assessing uncertainty in VGI for emergency response. Cartogr. Geogr. Inf. Sci. **41**, 440–455 (2014)
86. Foody, G.M., et al.: Accurate attribute mapping from volunteered geographic information: issues of volunteer quantity and quality. Cartogr. J. **52**, 336–344 (2015)
87. Goodhue, P., Delikostidis, I.: Modelling information quality and source reliability to improve the trust of volunteered geographic information (2017)
88. Vahidi, H., Klinkenberg, B., Yan, W.: Trust as a proxy indicator for intrinsic quality of volunteered geographic information in biodiversity monitoring programs. GISci. Remote Sens. **55**, 502–538 (2018)
89. Honarparvar, S., Malek, M.R., Saeedi, S., Liang, S.: Towards development of a real-time point feature quality assessment method for volunteered geographic information using the internet of things. ISPRS Int. J. Geo- Inf. **10**, 151 (2021)
90. Hou, Y., Biljecki, F.: A comprehensive framework for evaluating the quality of street view imagery. Int. J. Appl. Earth Obs. Geoinf. **115**, 103094 (2022)
91. Forati, A.M., Karimipour, F.: A VGI quality assessment method for VGI based on trustworthiness. GI Forum **4**, 3–11 (2016)
92. Dasgupta, A., Ghosh, S.K., Mitra, P.: A technique for assessing the quality of volunteered geographic information for disaster decision making. In: Gervasi, O., et al. (eds.) ICCSA 2018. LNCS, vol. 10960, pp. 589–597. Springer, Cham (2018). https://doi.org/10.1007/978-3-319-95162-1_40
93. Koswatte, S., McDougall, K., Liu, X.: VGI and crowdsourced data credibility analysis using spam email detection techniques. Int. J. Digit. Earth **11**, 520–532 (2018)
94. Ostermann, F.O., Spinsanti, L.: A conceptual workflow for automatically assessing the quality of volunteered geographic information for crisis management, vol. 2011, pp. 1–6 (2011)
95. Flanagin, A.J., Metzger, M.J.: The credibility of volunteered geographic information. GeoJournal **72**, 137–148 (2008)
96. Yanenko, O.: Volunteered geographic information and data quality-the case of social reporting (2015)

Author Index

B
Böhm, Klemens 27

C
Chbeir, Richard 114

E
El Ghazi, Aboubakr Achraf 27

F
Faiz, Sami 114
Friedl, Sabrina 68

M
Martinez-Gil, Jorge 99

N
Nciri, Donia 114

P
Pernul, Günther 68

S
Sassi, Salma 114
Sellami, Sana 1
Suntaxi, Gabriela 27

Y
Yin, Shaoyi 99

SPRINGER NATURE

GPSR Compliance

The European Union's (EU) General Product Safety Regulation (GPSR) is a set of rules that requires consumer products to be safe and our obligations to ensure this.

If you have any concerns about our products, you can contact us on ProductSafety@springernature.com

In case Publisher is established outside the EU, the EU authorized representative is:

Springer Nature Customer Service Center GmbH
Europaplatz 3
69115 Heidelberg, Germany

The manufacturer's authorised representative in the EU is Springer Nature Customer Service Centre GmbH, Europaplatz 3, 69115 Heidelberg, Germany. If you have any concerns regarding our products, please contact ProductSafety@springernature.com

Printed and bound by CPI Group (UK) Ltd, Croydon, CR0 4YY
25/03/2026
02078195-0013